普通高等学校"十四五"规划专业英语新形态精品教材

Ji-Dian Gongcheng Yingyu Yuedu

机电工程英语阅读

主编 ◎ 李迟　冷晟　　编者 ◎ 顾梦宇　钱菁

华中科技大学出版社
http://press.hust.edu.cn
中国·武汉

内 容 简 介

教材全面覆盖机电工程专业的各个重要领域,每个单元都根据大学英语教学要求和中国学生的英语学习特点精心设计,确保学生能够掌握所需的语言技能。内容包括:机械工程、生态机电一体化、生物机电一体化、电机工程学、材料科学、系统工程、机器人学、机电一体化和控制理论。每个单元都有 Text A 和 Test B 两篇课文:Text A 主要介绍围绕该研究方向的专业基本知识体系,课后练习的内容以符合《大学英语课程教学要求》为原则,围绕英语语言学综合知识体系进行设计;Test B 主要介绍某一专业方向上的研究成果,练习设计的主要目的是在不忽略四、六级硬性指标的基础上,帮助学生理解语言学知识并掌握机电工程与制造专业领域中的英语文本特点。

图书在版编目(CIP)数据

机电工程英语阅读 / 李迟,冷晟主编. -- 武汉:华中科技大学出版社,2024.5. --(普通高等学校"十四五"规划专业英语新形态精品教材). -- ISBN 978-7-5772-0932-6

Ⅰ. TH

中国国家版本馆 CIP 数据核字第 2024DB9713 号

机电工程英语阅读　　　　　　　　　　　　　　　　　　李　迟　冷　晟　主编
Ji-Dian Gongcheng Yingyu Yuedu

策划编辑:宋　焱
责任编辑:刘　凯
封面设计:廖亚萍
责任校对:张汇娟
责任监印:周治超

出版发行:华中科技大学出版社(中国·武汉)　　　电话:(027)81321913
　　　　　武汉市东湖新技术开发区华工科技园　　　邮编:430223
录　　排:华中科技大学出版社美编室
印　　刷:武汉开心印印刷有限公司
开　　本:787mm×1092mm　1/16
印　　张:13.5
字　　数:402千字
版　　次:2024年5月第1版第1次印刷
定　　价:55.00元

本书若有印装质量问题,请向出版社营销中心调换
全国免费服务热线:400-6679-118　　竭诚为您服务
版权所有　　侵权必究

引　言

随着国家"中国制造2025"和教育部"卓越工程师教育培养计划"的提出,机电工程与制造向着创新化、优质化、绿色化、人才化的方向高速发展,制造大国向着制造强国的目标逐步迈进,传统机电工程与制造行业逐渐掀起改革创新的热潮。而行业的发展离不开人才的支撑,"中国制造2025"和"卓越工程师教育培养计划"也无疑对人才的培养提出更高的要求。

在全球经济一体化的今天,英语成为机电制造业技术交流、科技合作的重要媒介,对具备熟练的英语交流能力的全面复合型制造业专业人才的需求更是与日俱增。专业英语是从业人员提升必备专业素质,了解国际前沿技术的重要途径。大学英语作为高校的一门基础课程,在"中国制造2025"和"卓越工程师教育培养计划"的背景下,必须做出相应的改革,改变过去英语教学的传统,进入用英语学、在用中学、在学中用的阶段,真正提高英语的综合应用能力。

专业英语是与某种特定专业相关的英语,是根据学习者特定的目的和需求开设的英语课程,其目的是培养学生在一定的工作环境中运用英语进行专业方面的工作与交流的一种能力。理工科的专业知识对于从事大学英语教学的教师来说是一道难题,机电方面知识的积累也并非靠一腔热血和一日之功就可以完成的。同时,机电工程与制造的专业英语教师多为机械专业出身,具备扎实的机械基础和深厚的专业研究功底,却没有经过完整、系统的英语语言学知识培训,在英语语言学教学方面与英语专业出身的教师还存在一定的差距。相关教师即使有国外留学经验,但缺乏扎实的语言学基础,也导致专业英语教学可能局限于书本上的词汇积累、句子翻译和英语论文的语法修改。所以,英语专业知识和机电工程与制造专业知识有机融合是大学英语改革的方向与出路,也是英语教师应该思考的问题。

这本教材并非传统意义上的机电工程与制造方面的专业英语教材,而是大学英语基础阶段教材。教材设计的理论基础是融合法教学理论与内容依托教学理论的有机结合,设计理念是用机电工程与制造有关的专业英语文本教授英语语言知识。这既弥补了大学英语教师在机电工程与制造方面的知识短板,也弥补了机电工程与制造专业英语教师的语言学知识的不足。通过这门课程的学习,学生可以在掌握机电专业基础的专业词汇表达、了解本专业的基础知识和发展趋势的同时,也具备基本的本专业的英语语言学素养,掌握一定的科技论文阅读、撰写能力。

本教材的另一个特点是融合了课程思政。它汇聚了"中国制造 2025"和"卓越工程师教育培养计划"的重要意义,立足于我国经济发展方式的转变需求,以创新为动力,以智能为引领,以绿色为催化,在先进制造、高端装备等关键领域引领制造业转型升级的浪潮。在改革战略中,"提升制造业国际化水平"被确立为主要任务和重点。这些都是以人才为核心的。学生深入了解机电工程与制造专业在国家战略中的重要地位之后,会增强专业认同感,更具时代感,其学习目标也更清晰,能够激发学习热情。

　　教材内容分为 9 个单元,每个单元有 Text A 和 Text B 两篇课文,Text A 以机电工程与制造的各个方面的基础知识为主,Text B 引入一些反映学科专业发展最新动态和最新技术的期刊论文、专著,培养和提高学生的学习兴趣,为学生将来走上科研之路打下良好的基础。课后练习的设计以测试为手段,以掌握语言学基本知识和高效检索信息为目的。教材在选材过程中,邀请了机电工程与制造方面的专家和教师共同进行选材的研讨,也请了教学经验丰富的英语教师从教学、测试和语言学的角度对体例进行设计,同时还邀请了几位机电工程与智能制造专业的同学对文本进行预读和专业词汇的整理。

编　者
2023 年 3 月 30 日
于南京航空航天大学天目湖校区

CONTENTS

Unit One Mechanical Engineering ... 1
  Text A The Evolution of Mechanical Engineering from Ancient Innovations to Modern Advancements ... 2
  Text B Advancements in DFMA Methodologies ... 15

Unit Two Ecomechatronics ... 27
  Text A Advancing Sustainable Machine Design ... 28
  Text B Supervised Learning: ML Model Development Basics ... 39

Unit Three Biomechatronics ... 51
  Text A Biomechatronics: Bridging Biology and Engineering ... 52
  Text B Minimally Invasive Surgery and Surgical Robotics ... 63

Unit Four Electromechanics ... 75
  Text A Evolution and Applications of Electromechanics ... 76
  Text B Core Principles for Flexible Surgical Robotics in MIS ... 87

Unit Five Materials Science ... 98
  Text A Materials Science: Interdisciplinary Research and Applications ... 99
  Text B Material Structure: From Atoms to Macroscopic Scale ... 109

Unit Six Systems Engineering ... 120
  Text A Holistic Approaches for Complex Systems ... 121
  Text B Navigating Complex Systems ... 131

Unit Seven Robotics ... 143
  Text A The Transformative Field of Robotics ... 144
  Text B Autonomous Modular Robot System Tasks, Adaptability, and Future Improvements ... 153

Unit Eight Mechatronics ··· 164
Text A The Versatility of Mechatronics Engineering ······························ 165
Text B Intelligent Manufacturing Execution Systems ····························· 176

Unit Nine Control Theory ·· 186
Text A Evolving Control Theory: From Engineering to AI ···················· 187
Text B Closed-Loop Control: Achieving Desired System Output ············ 196

Unit One　Mechanical Engineering

Warm-up

On May 8, 2015, China's State Council unveiled its first 10-year national plan for transforming China's manufacturing, entitled "Made in China 2025". The plan is designed to put China on a new path to industrialization, with greater emphasis on innovation, expanded use of new-generation information technology, intelligent manufacturing, consolidation of the industrial base, integration of industrial process and systems, and a robust multilayer talent development structure. Measures taken in this respect will facilitate China's transformation from a manufacturing giant with a focus on quantity to one with a qualitative edge.

Thought-provoking Questions

· How can the concepts of innovation and intelligent manufacturing be effectively integrated into the traditional manufacturing processes in China? What role can mechanical engineering students play in driving this transformation?

· What steps can be taken to establish a robust multilayer talent development structure in the field of mechanical manufacturing in China? How can mechanical engineering students contribute to nurturing and developing the required skill sets?

Text A
The Evolution of Mechanical Engineering from Ancient Innovations to Modern Advancements

Research Background

Mechanical engineering is an engineering branch that focuses on the study of physical machines involving force and movement. It combines principles of engineering physics, mathematics, and materials science to design, analyze, manufacture, and maintain mechanical systems. With its roots dating back several thousand years, mechanical engineering emerged as a distinct field during the Industrial Revolution in Europe. It has since evolved and expanded to encompass various disciplines, including aerospace engineering, metallurgical engineering, and biomedical engineering. Throughout history, mechanical engineering has seen significant contributions from ancient civilizations, Islamic inventors, and notable figures such as Archimedes and Isaac Newton. The development of machine tools during the 19th-century Industrial Revolution further propelled the field, leading to the formation of professional societies and engineering education institutions worldwide.

Introduction Questions

· What are some key areas of study within mechanical engineering education?
· How did the Industrial Revolution contribute to the development of mechanical engineering as a separate field?
· What were some significant milestones in the history of mechanical engineering education in the United States?

A. Mechanical engineering is the study of physical machines that may involve force and movement. It is an engineering branch that combines engineering physics and mathematics principles with materials science, to design, analyze, manufacture, and maintain mechanical systems. It is one of the oldest and broadest of the engineering branches. Mechanical engineering requires an understanding of core areas including mechanics, dynamics, thermodynamics, materials science, design, structural analysis, and electricity. In addition to these core principles, mechanical engineers use tools such as computer-aided design (CAD), computer-aided manufacturing (CAM), and product lifecycle management to design and analyze manufacturing plants, industrial equipment and machinery, heating and cooling systems, transport systems, aircraft, watercraft, robotics, medical devices, weapons, and others.

B. Mechanical engineering emerged as a field during the Industrial Revolution in Europe in the 18th century; however, its development can be traced back several thousand years around the world. In the 19th century, developments in physics led to the development of mechanical engineering science. The field has continually evolved to incorporate advancements; today mechanical engineers are pursuing developments in such areas as composites, mechatronics, and nanotechnology. It also overlaps with aerospace engineering, metallurgical engineering, civil engineering, structural engineering, electrical engineering, manufacturing engineering, chemical engineering, industrial engineering, and other engineering disciplines to varying amounts. Mechanical engineers may also work in the field of biomedical engineering, specifically with biomechanics, transport phenomena, biomechatronics, bionanotechnology, and modeling of biological systems.

C. The application of mechanical engineering can be seen in the archives of various ancient and medieval societies. The six classic simple machines were known in the ancient Near East. The wedge and the inclined plane (ramp) were known since prehistoric times. The wheel, along with the wheel and axle mechanism, was invented in Mesopotamia (modern Iraq) during the 5th millennium BC. The lever mechanism first appeared around 5,000 years ago in the Near East, where it was used in a simple balance scale, and to move large objects in ancient Egyptian technology. The lever was also used in the shadoof water-lifting device, the first crane machine, which appeared in Mesopotamia circa 3000 BC. The earliest evidence of pulleys date back to Mesopotamia in the early 2nd millennium BC.

D. The sakia was developed in the Kingdom of Kush during the 4th century BC. It relied on animal power reducing the tow on the requirement of human energy. Reservoirs in the form of hafirs were developed in Kush to store water and boost irrigation. Bloomeries and blast furnaces were developed during the seventh century BC in Meroe. Kushite sundials applied mathematics in the form of advanced trigonometry.

E. The earliest practical water-powered machines, the water wheel and watermill, first appeared in the Persian Empire, in what are now Iraq and Iran, by the early 4th century BC. In ancient Greece, the works of Archimedes (287-212 BC) influenced mechanics in the Western tradition. In Roman Egypt, Heron of Alexandria (c. 10-70 AD) created the first steam-powered device (Aeolipile). In China, Zhang Heng (78-139 AD) improved a water clock and invented a seismometer, and Ma Jun (200-265 AD) invented a chariot with differential gears. The medieval Chinese horologist and engineer Su Song (1020-1101 AD) incorporated an escapement mechanism into his astronomical clock tower two centuries before escapement devices were found in medieval European clocks. He also invented the world's first known endless power-transmitting chain drive.

F. During the Islamic Golden Age (7th to 15th century), Islamic inventors made remarkable contributions in the field of mechanical technology. Al-Jazari, who was one of them, wrote his famous *Book of Knowledge of Ingenious Mechanical Devices* in 1206 and presented many mechanical designs.

G. In the 17th century, important breakthroughs in the foundations of mechanical engineering occurred in England and the Continent. The Dutch mathematician and physicist Christiaan Huygens invented the pendulum clock in 1657, which was the first reliable timekeeper for almost 300 years, and published a work dedicated to clock designs and the theory behind them. In England, Isaac Newton formulated Newton's Laws of Motion and developed the calculus, which would become the mathematical basis of physics. Newton was reluctant to publish his works for years, but he was finally persuaded to do so by his colleagues, such as Edmond Halley. Gottfried Wilhelm Leibniz, who earlier designed a mechanical calculator, is also credited with developing the calculus during the same time period.

H. During the early 19th century Industrial Revolution, machine tools were developed in England, Germany, and Scotland. This allowed mechanical engineering to develop as a separate field within engineering. They brought with them manufacturing machines and the engines to power them. The first British professional society of mechanical engineers was formed in 1847—Institution of Mechanical Engineers, thirty years after the civil engineers formed the first such professional society—Institution of Civil Engineers. On the European continent, Johann von Zimmermann (1820-1901) founded the first factory for grinding machines in Chemnitz, Germany in 1848.

I. In the United States, the American Society of Mechanical Engineers (ASME) was formed in 1880, becoming the third such professional engineering society, after the American Society of Civil Engineers (1852) and the American Institute of Mining Engineers (1871). The first schools in the United States to offer an engineering education were the United States Military Academy in 1817, an institution now known as Norwich University in 1819, and Rensselaer Polytechnic Institute in 1825. Education in mechanical engineering has historically been based on a strong foundation in mathematics and science.

Vocabulary List

mechanical engineering
the branch of engineering that deals with the design, construction, and operation of machinery 机械工程

aerospace
the branch of technology and industry concerned with aviation and space flight 航空航天

metallurgical engineering
the branch of engineering that deals with the science and technology of metals 冶金工程

biomedical
relating to the application of engineering principles to medicine and biology 生物医学

biomedical engineering
the application of engineering principles and techniques to the medical and healthcare field 生物医学工程

civil engineering
the branch of engineering that deals with the design and construction of public works such as bridges, roads, and buildings 土木工程

dynamics
the branch of mechanics that deals with the motion of objects and the forces causing the motion 动力学

Mesopotamia
an ancient region located in the eastern Mediterranean, primarily corresponding to modern-day Iraq 美索不达米亚

Meroe

an ancient city in present-day Sudan 美洛伊

thermodynamics

the branch of physics concerned with the relationships between heat and other forms of energy 热力学

nanotechnology

the science, engineering, and application of materials and devices with structures and properties that exist at the nanometer scale 纳米技术

biomechatronics

the integration of mechanical components with electronics and biological systems 生物机电一体化

bionanotechnology

the application of nanotechnology to biological systems 生物纳米技术

sakia

a type of water wheel used for irrigation or milling 水车

Kushite sundials

sundials used by the Kushite civilization 库施典阳历仪器

Language Enhancement

Ⅰ. Complete the following sentences with words listed in the box below. Change the form where necessary.

combine	incorporate	ingenious	grind	propel
elaborate	trace	boost	encompass	composite
innovation	reluctant	formulate	distinct	pursue

1. Newton's Laws of Motion, _____ in the 17th century, laid the foundation for understanding the principles of mechanical engineering.

2. Mechanical engineering _____ various disciplines such as mathematics, physics, and materials science to design and analyze mechanical systems.

3. Al-Jazari's Book of Knowledge of _____ Mechanical Devices showcased his innovative designs that were ahead of their time.

4. Mechanical engineering is a distinct field that _____ a wide range of applications, from aerospace engineering to biomedical engineering.

5. The Industrial Revolution brought significant _____ that propelled the development of mechanical engineering as a separate discipline.

6. Reservoirs were constructed in Kush to _____ irrigation, highlighting the ancient engineering techniques used to enhance water supply.

7. Mechanical engineers constantly _____ new ideas and innovations to improve existing systems and develop novel technologies.

8. The engineers had to _____ on their designs, providing detailed specifications and calculations for the mechanical systems they were developing.

9. _____, such as carbon fiber reinforced polymers, are increasingly used in mechanical engineering applications for their high strength-to-weight ratio.

10. Isaac Newton was initially _____ to publish his works but eventually shared his groundbreaking theories on motion and calculus.

11. Mechanical engineering _____ theoretical knowledge with practical skills to grind and shape materials into precise components.

12. The history of mechanical engineering can be _____ back thousands of years, demonstrating its long-standing presence in human civilization.

13. The _____ feature of this new machine design is its compact size, making it ideal for tight spaces in industrial settings.

14. The jet engine's powerful thrust can _____ the aircraft forward at incredible speeds.

15. The machine uses abrasive wheels to _____ metal parts to precise specifications in the manufacturing process.

Ⅱ. **Complete the following sentences with phrases listed in the box below. Change the form where necessary.**

be reluctant to	base on	be credited with	overlap with
be known as	in the form of	combine with	in addition to
lead to	rely on	traced back	in the field of

1. The new technology _____ artificial intelligence to enhance efficiency in manufacturing processes.

2. _____ improving communication, the app also offers a range of additional features.

3. The origins of the coding language can _____ the 1950s.

4. The use of outdated software can _____ compatibility issues and security vulnerabilities.

5. The responsibilities of the two departments _____ each other, causing confusion among employees.

6. _____ robotics, significant advancements have been made in recent years.

7. Many businesses _____ cloud computing services for their data storage and processing needs.

8. The use of blockchain technology provides security and transparency _____ decentralized networks.

9. Some individuals may _____ adopt new technologies due to privacy concerns.

10. The renowned scientist _____ inventing the first electric car.

11. Elon Musk _____ a visionary entrepreneur in the field of space exploration.

12. The development of the new software _____ user feedback and market research.

Ⅲ. Complete the summary with words listed in the box. Change the form where necessary.

maintain	include	overlap	medieval	movement
composite	trace	involve	discipline	evolve
physical	emerge	model	lifecycle	combine

Mechanical engineering 1)_____ the study and application of 2)_____ machines that use force and 3)_____. This field 4)_____ engineering physics, mathematics, and materials science to design, movement, manufacture, and 5)_____ mechanical systems. Mechanical engineers must understand mechanics, dynamics, thermodynamics, materials science, design, structural analysis, and electricity. They use tools such as CAD, CAM, and product 6)_____ management to design and analyze various systems, 7)_____ manufacturing plants, industrial equipment, transport, medical devices, robotics, and more. The history of mechanical engineering can be 8)_____ back thousands of years through various ancient and 9)_____ societies, with inventions such as the water wheel, pulleys, and the lever. During the Industrial Revolution in the 18th century, mechanical engineering 10)_____ as a field, and it continues to 11)_____ today with advancements in 12)_____, mechatronics, and nanotechnology. Mechanical engineering 13)_____ with other engineering 14)_____, and mechanical engineers can work in biomedical engineering with biomechanics, transport phenomena, biomechatronics, bionanotechnology, and 15)_____ of biological systems.

Academic Expression

Pair work: Discuss with your partner and compare the two possible paraphrases of each sentence and decide which one expresses the original meaning more academically.

1. Mechanical engineering requires an understanding of core areas including mechanics, dynamics, thermodynamics, materials science, design, structural analysis, and electricity.

    a. Mechanical engineering necessitates a comprehensive grasp of fundamental disciplines, encompassing mechanics, dynamics, thermodynamics, materials science, design, structural analysis, and electricity.

    b. To be a mechanical engineer, you gotta know your stuff in mechanics, dynamics, thermodynamics, materials science, design, structural analysis, and electricity.

2. The field has continually evolved to incorporate advancements; today mechanical engineers are pursuing developments in such areas as composites, mechatronics, and nanotechnology.

    a. The field of mechanical engineering has been evolving constantly, keeping up with all the cool new tech. Nowadays, mechanical engineers are all about composites, mechatronics, and nanotechnology.

    b. The field has undergone continuous evolution to assimilate technological advancements; presently, mechanical engineers are actively engaged in pioneering endeavors in domains like composites, mechatronics, and nanotechnology.

3. Mechanical engineers may also work in the field of biomedical engineering, specifically with biomechanics, transport phenomena, biomechatronics, bionanotechnology, and modeling of biological systems.

    a. Mechanical engineers may also find professional opportunities in the realm of biomedical engineering, particularly specializing in biomechanics, transport phenomena, biomechatronics, bionanotechnology, and the modeling of biological systems.

    b. Mechanical engineers can also work in the field of biomedical engineering, doing cool stuff like studying how bodies move (biomechanics), figuring out how things get transported (transport phenomena), creating bionic gadgets (biomechatronics), playing around with tiny particles (bionanotechnology), and modeling biological systems.

4. The earliest practical water-powered machines, the water wheel and watermill, first appeared in the Persian Empire, in what are now Iraq and Iran, by the early 4th century BC.

    a. You won't believe it, but the very first water-powered machines like the water wheel and watermill were rocking it in the Persian Empire back in the early 4th century BC, in what we now call Iraq and Iran.

    b. The earliest instances of practical water-powered machinery, namely the water wheel and watermill, made their debut in the Persian Empire, present-day Iraq and Iran, during the early 4th century BC.

5. The lever mechanism first appeared around 5,000 years ago in the Near East, where it was used in a simple balance scale, and to move large objects in ancient Egyptian technology.

    a. The lever mechanism, tracing back 5,000 years ago to the Near East, played a pivotal role in the operation of simple balance scales and the movement of large objects in ancient Egyptian technology.

    b. Check this out: the lever mechanism has been around for a whopping 5,000 years! They used it back in the day in the Near East for simple balance scales and moving big objects in ancient Egyptian technology.

6. The medieval Chinese horologist and engineer Su Song (1020-1101 AD) incorporated an escapement mechanism into his astronomical clock tower two centuries before escapement devices were found in medieval European clocks.

    a. Su Song, a genius Chinese dude from way back in 1020-1101 AD, was ahead of his time. He put an escapement mechanism in his astronomical clock tower a couple of centuries before Europeans got the memo.

    b. Su Song, a distinguished Chinese horologist and engineer of the medieval era (1020-1101 AD), pioneered the integration of an escapement mechanism into his astronomical clock tower, pre-dating its appearance in medieval European clocks by two centuries.

7. In the 17th century, important breakthroughs in the foundations of mechanical engineering occurred in England and the Continent.

    a. The 17th century witnessed significant breakthroughs in the foundational principles of mechanical engineering, with notable advancements originating from England and the Continent.

b. Back in the 17th century, mechanical engineering was going through some major changes. England and the Continent were where the action was happening, with groundbreaking advancements left and right.

8. In the United States, the American Society of Mechanical Engineers (ASME) was formed in 1880, becoming the third such professional engineering society, after the American Society of Civil Engineers (1852) and the American Institute of Mining Engineers (1871).

   a. The American Society of Mechanical Engineers (ASME) came onto the scene in 1880, making it the third big professional engineering society in the US. Those guys followed in the footsteps of the American Society of Civil Engineers (1852) and the American Institute of Mining Engineers (1871).

   b. In 1880, the formation of the American Society of Mechanical Engineers (ASME) marked the establishment of the third professional engineering society in the United States, following the American Society of Civil Engineers (1852) and the American Institute of Mining Engineers (1871).

Understanding the Text

Ⅰ. **Pair work: Work with your partner and take turns asking and answering the following questions according to the information contained in the text.**

1. What is mechanical engineering?

2. How old is the field of mechanical engineering?

3. What are some core areas of knowledge required in mechanical engineering?

4. What are some tools used by mechanical engineers in their work?

5. How does mechanical engineering overlap with other engineering disciplines?

6. What are some historical examples of mechanical engineering applications?

7. What were some notable developments during the Industrial Revolution related to mechanical engineering?

8. When did professional societies and educational institutions for mechanical engineering form?

_____

Ⅱ. **Choose the best answer according to Text A.**

1. What is mechanical engineering?

A) The study of physical machines that may involve force and movement.

B) The design of software systems for manufacturing plants and machinery.

C) The science of structural analysis and electricity.

D) The development of advanced materials science for the aviation industry.

2. What areas of study are required for mechanical engineering?

A) Only materials science and design.

B) Only electricity and structural analysis.

C) Mechanics, dynamics, thermodynamics, materials science, design, structural analysis, and electricity.

D) Only computer-aided design and computer-aided manufacturing.

3. What are some tools used by mechanical engineers?

A) Computer-aided design, computer-aided manufacturing, and product lifecycle management.

B) Computer programming, data analysis, and data visualization.

C) Graphic design software, video editing software, and animation software.

D) Social media platforms, email, and chat apps.

4. When did mechanical engineering emerge as a field?

A) During the Industrial Revolution in Europe in the 18th century.

B) In ancient Egypt.

C) During the Islamic Golden Age.

D) In the early 19th century Industrial Revolution.

5. What ancient societies contributed to the development of mechanical engineering?

A) Ancient Greece and Rome.

B) Ancient Egypt and Mesopotamia.

C) Ancient China and Persia.

D) All of the above.

6. Who invented the first steam-powered device?
A) Archimedes.
B) Heron of Alexandria.
C) Zhang Heng.
D) Isaac Newton.

7. When was the first British professional society of mechanical engineers formed?
A) 1847.
B) 1852.
C) 1871.
D) 1880.

8. What was the first reliable timekeeper invented by Christiaan Huygens?
A) The water clock.
B) The pendulum clock.
C) The sundial.
D) The hourglass.

III. Identify the paragraph from which the information is derived. You may use any letter more than once.

1. During the 4th century BC, the Kingdom of Kush introduced the sakia, a mechanism leveraging animal energy to minimize human exertion, and also devised reservoirs for water storage, facilitating irrigation.

2. Mechanical engineering intersects with an array of other engineering fields, such as aerospace engineering, civil engineering, electrical engineering, to varying extents.

3. Mechanical engineers employ some tools to facilitate the creation and assessment of diverse systems and machinery.

4. The advent of the Industrial Revolution in the early 19th century served as a catalyst for the advancement of mechanical engineering, introducing machine tools and giving rise to professional societies.

5. The Islamic Golden Age witnessed remarkable contributions to mechanical technology, featuring inventors like Al-Jazari, who showcased their inventive mechanical designs.

6. Mechanical engineering's evolution spans millennia, integrating progressions in disciplines like composites, mechatronics, and nanotechnology.

7. The field of mechanical engineering encompasses the examination of tangible machines that encompass force and movement, merging principles from engineering physics, mathematics, and materials science.

8. The utilization of mechanical engineering can be witnessed in the historical achievements of past societies, exemplified by the creation of rudimentary machines like the wheel, lever, and pulley.

9. The earliest instances of water-powered machines, like the water wheel and watermill, originated in the Persian Empire approximately in the 4th century BC.

10. The realm of mechanical engineering finds its application in various domains, encompassing the creation and upkeep of manufacturing facilities, industrial apparatus, transportation networks, medical instruments, and others.

Rise of a Great Power

Translate the following Chinese part into English and make it a complete English text.

SuSong (1020-1101), was an astronomer, astronomical machinery maker and pharmacologist of the Northern Song Dynasty. 1)（他领导制造了世界上最古老的天文钟"水运仪象台（water-driven astronomical clock tower）",开启近代钟表擒纵器（modern clock escapement）的先河）_____

_____.

For his outstanding contributions to science and technology, especially medicine and astronomy, 2)（被称为"中国古代和中世纪最伟大的博物学家和科学家之一"）_____

_____

_____. Joseph Needham, a British historian of science and technology, spoke highly of him: 3)（苏颂把时钟机械和观察用浑仪结合起来,在原理上已经完全成功。因此可以说他比罗伯特·胡克先行了六个世纪。）"

_____

_____"

The number of stars observed in Europe before the Renaissance in the 14th century was 1022, 422 fewer than Su's. Therefore, Western historians of science and technology such as Tiller, Brown and Sutton even believed that 4)（从中世纪直到14世纪末,除中国的星图以外,再也举不出别的星图了。）_____

_____.

Text B
Advancements in DFMA Methodologies

Research Background

DFMA (Design for Manufacturing and Assembly) is a category of methods aimed at optimizing the manufacturing and assembly phases of a product. In the early stages of DFMA method development, the focus was on conceptualization and description, with academic and exemplary case studies being provided. During the 1990s, there was an exponential increase in the application of DFMA methods in industries, particularly in the mechanical field. Subsequently, a series of case studies were conducted, demonstrating the applicability of DFMA in mechanical and electro-mechanical products. Most publications featuring case studies were implemented in industrial settings, using generic products with few components to validate the methods and ensure their reliability.

Introduction Questions

· How has the application of DFMA methods evolved over time, particularly in the mechanical field?

· What types of products have been analyzed using DFMA methods, and how have they varied in different fields, such as automotive and aerospace?

· What factors have contributed to the increased publication of DFMA methods for complex products in the last two decades, and why is the consideration of the whole system challenging in traditional DFMA?

A. The design for manufacturing and assembly (DFMA) is a family of methods belonging to the design for X (DFX) category which goal is to optimize the manufacturing and assembly phase of a product. At the beginning of DFMA method development (early 1980s), articles were focusing on the conceptualization and description of DFMA methods, providing academic and exemplary case studies. During the 1990s, the application of DFMA methods in industries increased exponentially, particularly in the mechanical field. Starting from the second decade (D2), several case studies were provided to demonstrate the applicability of DFMA in mechanical and electro-mechanical products, and the same trend was confirmed in the following decades (D3 and D4). It is worth noting that most of the publications giving case studies have been implemented in the industrial field. The reason lies in the fact that several DFMA methods available in the literature are tested on generic products made of few components (i. e., dust filters, stapler, boiler) to validate the methods and their reliability.

B. The number of papers presenting case studies in the automotive and aerospace fields is well balanced. Products analyzed with DFMA methods are varying from sub-assemblies of a car (i. e., the suspension system, brake and clutch) to aircraft systems (i. e., pilot instrument panel, contactor assembly). Only a few articles tried to tackle the assemblability of a whole product; among them, Thompson et al. tried to point out the relation between DFMA rules and late design changes in high-speed product development (i. e., circulator pumps for the commercial building services market). Gerding et al. tackles the problem of implementing DFMA rules in long-lead-time products (i. e., aircraft), while Barbosa and Carvalho proposed DFMA rules to optimize the assembly phase of an aircraft through re-design actions.

C. To understand the interest of the topic over time, the Publications' year was analyzed together with the type of publication (i. e., journal or conference proceeding). Papers describing DFMA applications on both complex and simple products have increased over the years. It is interesting to notice that most of the articles proposing DFMA methods for complex products have been published in the last two decades (D3 and D4). This trend may be justified by several reasons. The first one concerns the fact that more and more industries are focusing on reaching a global improvement of their product, making the application of traditional DFMA challenging since the whole system must be considered. Another major factor in the development of DFMA methods for complex products concerns the increment of processing power that allows designers and engineers to handle a high amount of data in a limited time-frame, widening the boundary of their optimization problem from sub-parts to the whole system.

D. The study of DFMA methods applied to simple products in the last three decades has increased as well. However, for the last decade (D4) most of the papers are published in conference proceedings and they present applications of already well-known DFMA techniques on different systems. Despite these works being useful to increase the number of case studies where DFMA methods are applied, they cannot be considered as research advancement in the DFMA methods.

E. Other works published in conference proceedings are trying to extend DFMA principles in several ways. For example, Esterman and Kamath attempted to apply DFMA to the improvement of assembly lines, Wood et al. and Nyemba et al. provided new design rules to cope with constraint production of the developing countries, and finally Favi et al., Hein et al., and Gupta and Kumar included new principles and criteria for multi-objective analysis (i.e., cost, sustainability).

F. From the performed analysis, DFMA methods have been mainly applied on simple products or sub-assemblies, in which all parts are made with traditional production technologies (i.e., fusion, sheet metal stamping and bending, forging). DFMA analysis evaluates assembly solutions adopted in the analyzed products. Assembly solutions are generally bolted joints, more rarely welded or riveted joints. The main goal of these analyses is to understand if it is possible to reduce the number of components which, typically, leads to a reduction of assembly time. As an outcome, the typical product analyzed using DFMA techniques is a simple product assembled manually with bolted joints made of less than 60 parts. Another interesting result concerns the fact that sub-assemblies are considered rather than the whole product. This result leads to the application of DFMA methodologies in a limited context (i.e., the companies which are designing and manufacturing sub-assemblies) making effective the benefits of DFMA for suppliers.

G. In this scenario, each module (sub-assembly) is assembled with a specific assembly technology, making the overall analysis easier to manage. For instance, a car engine is assembled with bolted joints and chassis are assembled with welding technologies. If the assembly technology varies, then the DFMA analysis becomes more challenging and, consequently, the overall final improvement might not have an elevated positive impact as the sub-systems improvements might have.

Vocabulary List

conceptualization
the process of forming a concept 概念化

exemplary
serving as a model or an example 典范的,榜样的

exponential

rapidly increasing 指数增长的

electro-mechanical

involving both electrical and mechanical aspects 电气机械的

suspension

the system of springs, shock absorbers, and linkages that connects a vehicle to its wheels 悬挂系统

clutch

a device for connecting and disconnecting two rotating shafts 离合器

contactor assembly

a unit composed of contactors 接触器组装

metal stamping

a manufacturing process of shaping metal sheets 金属冲压

bolted joint

a connection made using bolts 螺栓连接

Language Enhancement

Ⅰ. Complete the following sentences with words listed in the box below. Change the form where necessary.

validate	reliability	constraint	forge	tackle
elevate	demonstrate	optimize	assemble	extend
evaluate	attempt	subsequently	justify	confirm

1. To _____ the productivity of our assembly line, we implemented robotic arms to handle repetitive tasks.

2. The engineer made a(n) _____ to streamline the manufacturing process by introducing a new automation system.

3. We _____ the production hours to meet the high demand for the product.

4. The data analysis _____ the need for additional resources in order to meet the project deadline.

5. The quality control team _____ that all components met the required specifications before proceeding with assembly.

6. Rigorous testing was conducted to _____ the performance and reliability of the newly developed machine.

7. The _____ of the prototype showcased its advanced features and capabilities to potential investors.

8. _____, the parts were carefully assembled to create the final product.

9. Time constraint posed a challenge, but the team worked together to _____ the issue and deliver on time.

10. The engineers _____ the efficiency of the production line and identified areas for improvement.

11. The _____ of the machine was enhanced through continuous monitoring and maintenance.

12. The _____ process aimed to reduce waste and maximize productivity.

13. The company _____ strategic partnerships with suppliers to ensure a steady supply of high-quality materials.

14. To tackle complex _____ tasks, the workers received specialized training and guidance.

15. The project faced various _____, such as budget limitations and time constraints, but the team managed to overcome them through effective planning and resource allocation.

II. Complete the following sentences with phrases listed in the box below. Change the form where necessary.

attempt to	cope with	point out	vary from	consider as
be made of	at the beginning of	lie in	focus on	belonging to

1. The new smartphone _____ the latest series offers cutting-edge features and improved performance.

2. _____ the coding workshop, the instructor introduced the basic concepts of programming.

3. The research project _____ developing renewable energy sources for sustainable power generation.

4. The key to understanding quantum mechanics _____ studying the behavior of subatomic particles.

5. The processing time for different algorithms can _____ a few milliseconds to several hours.

6. The professor _____ the significance of data privacy in the digital age during the lecture.

7. The new invention _____ a breakthrough in medical technology.

8. The scientists _____ create a synthetic material with superior strength and durability.

9. With the rapid advancements in technology, businesses need to find ways _____ the ever-changing market demands.

10. The sculpture in the art museum _____ bronze, showcasing the artist's skill and creativity.

Ⅲ. Complete the summary with words listed in the box. Change the form where necessary.

extend	constraint	proceed	optimize	conceptualization
emphasis	engage	manufacturing	exponential	application
diverse	assembly	covering	methodology	evaluate

DFMA (design for manufacturing and assembly) is a method aimed at 1)_____ the 2)_____ and 3)_____ phase of products. Initially focused on 4)_____ and case studies, DFMA methods gained 5)_____ popularity in the mechanical field during the 1990s. The automotive and aerospace industries witnessed a balanced number of case studies, 6)_____ various components and systems. Over time, there has been a rise in papers describing DFMA 7)_____ for both complex and simple products, with a particular 8)_____ on complex products in the last two decades. DFMA 9)_____ have found success in sub-assembly analysis, 10)_____ assembly solutions and reducing components and assembly time, primarily through bolted joints. Conference 11)_____ have 12)_____ DFMA principles, exploring applications in assembly line improvement, design for production 13)_____, and multi-objective analysis. While DFMA methods have been effective for suppliers 14)_____ in sub-assembly manufacturing, there is a growing need to address the challenges posed by varying assembly technologies. Overall, DFMA offers promising opportunities for optimizing manufacturing and assembly processes in 15)_____ industries.

Unit One Mechanical Engineering

Academic Expression

Pair work: Discuss with your partner and compare the two possible paraphrases of each sentence and decide which one expresses the original meaning more academically.

1. The design for manufacturing and assembly (DFMA) is a family of methods belonging to the design for X (DFX) category which goal is to optimize the manufacturing and assembly phase of a product.

  a. DFMA is a set of methods that falls under the design for X (DFX) category. Its main objective is to optimize the manufacturing and assembly process of a product.

  b. The DFMA methodology, categorized under design for X (DFX), encompasses a set of methods aimed at optimizing the manufacturing and assembly stage of a product.

2. Papers describing DFMA applications on both complex and simple products have increased over the years.

  a. The number of publications documenting the application of DFMA on products of varying complexity has witnessed a steady increase over time.

  b. More and more papers are being published that describe how DFMA is used on both complex and simple products.

3. It is interesting to notice that most of the articles proposing DFMA methods for complex products have been published in the last two decades.

  a. It's worth noting that most articles suggesting DFMA methods for complex products have been published in the last two decades.

  b. An intriguing observation is that the majority of articles proposing DFMA methods for complex products have been published in the past two decades.

4. The study of DFMA methods applied to simple products in the last three decades has increased as well.

  a. The research on the application of DFMA methods to simple products has also experienced a notable growth over the past three decades.

  b. The study of DFMA methods applied to simple products has also increased significantly in the last three decades.

5. Despite these works being useful to increase the number of case studies where DFMA

methods are applied, they cannot be considered as research advancement in the DFMA methods.

    a. While these works help increase the number of case studies using DFMA methods, they shouldn't be seen as major breakthroughs in DFMA research.

    b. Although these works contribute to the expansion of case studies applying DFMA methods, they should not be regarded as significant research advancements in the field of DFMA methods.

6. For example, Esterman and Kamath attempted to apply DFMA to the improvement of assembly lines.

    a. As an illustration, Esterman and Kamath made efforts to apply DFMA principles to enhance assembly line operations.

    b. For instance, Esterman and Kamath tried to use DFMA to improve assembly lines.

7. DFMA analysis evaluates assembly solutions adopted in the analyzed products.

    a. DFMA analysis looks at the assembly solutions used in the products under study.

    b. DFMA analysis assesses the assembly solutions implemented in the examined products.

8. If the assembly technology varies, then the DFMA analysis becomes more challenging and, consequently, the overall final improvement might not have an elevated positive impact as the sub-systems improvements might have.

    a. When the assembly technology differs, the DFMA analysis becomes more complex, and as a result, the overall final improvement may not have as significant of a positive impact as the improvements made to the sub-systems.

    b. If the assembly methods vary, it becomes more challenging to analyze using DFMA, and therefore, the overall improvement may not have as big of a positive impact as the improvements made to the individual parts.

Understanding the Text

Ⅰ. **Text B has seven paragraphs, A-G. Which paragraph contains the following information? You may use any letter more than once.**

1. Varying assembly technologies pose challenges, and improvements at the sub-system

level may have a more significant impact than overall improvements.

2. DFMA methods have been primarily tested on generic industrial products with few components to ensure their reliability.

3. There has been an increase in papers describing the application of DFMA methods for both complex and simple products, with a notable surge in articles proposing methods for complex products in the last two decades.

4. DFMA methodologies are particularly effective for suppliers engaged in designing and manufacturing sub-assemblies, as they analyze and optimize at that level instead of the entire product.

5. Case studies in the automotive and aerospace fields using DFMA methods cover a wide range of components, including car sub-assemblies like the suspension system and aircraft systems such as the pilot instrument panel.

6. DFMA methods primarily focus on simple products or sub-assemblies using traditional production technologies like fusion, stamping, bending, and forging.

7. Conference proceedings include works that have expanded DFMA principles, such as applying them to enhance assembly lines, introducing design rules for production constraints in developing countries, and incorporating multi-objective analysis criteria.

8. Each module or sub-assembly is assembled using specific technology, facilitating the overall DFMA analysis.

9. In the past decade, most papers on DFMA methods for simple products were published in conference proceedings, featuring applications of well-established techniques on different systems.

10. Evaluation of assembly solutions, particularly bolted joints, aims to reduce components and assembly time.

II. Do the following statements agree with the information given in Text B? Write your judgment.

TRUE   if the statement agrees with the information
FALSE   if the statement contradicts the information
NOT GIVEN   if there is no information on this

1. The goal of the design for manufacturing and assembly (DFMA) is to optimize the manufacturing and assembly phase of a product.

2. During the 1990s, the application of DFMA methods in industries increased exponentially, particularly in the manufacturing field.

3. The number of papers presenting case studies in the automotive and aerospace fields is well fruitful.

4. Thompson et al. tried to point out the relation between DFMA rules and late design changes in high-speed product development.

5. Most of the articles proposing DFMA methods for complex products have been published in the last two decades (D3 and D4).

6. The study of DFMA methods applied to various products has increased as well.

7. Esterman and Kamath attempted to apply DFMA to the improvement of assembly lines.

8. DFMA analysis evaluates assembly solutions adopted in the analyzed products, and the main goal is to understand if it is possible to reduce the number of components.

9. The typical product analyzed using DFMA techniques is a simple product assembled manually with bolted joints made of fewer than 80 parts.

10. Each module (sub-assembly) is assembled with a specific assembly technology, making the overall analysis easier to manage.

III. Group work: Please work in groups and complete the structural table with the information from Text B.

Function	Paragraph	Content
Introduction	A	Discusses its belonging to the design for X (DFX) category and its goal of optimizing 1) _____ and assembly. It mentions the early focus on 2) _____ and case studies.
Body	B	Presents case studies in the 3) _____ and aerospace fields, highlighting the range of products analyzed with DFMA methods.

Continued Table

Function	Paragraph	Content
Body	C	Analyzes the trend of DFMA publications over time, with a focus on complex products in the last two decades. It 4)_____ the reasons behind this trend, including the increasing focus on global product 5)_____ and the advancements in processing power.
	D	Discusses the increase in the study of DFMA applied to simple products but notes the lack of research 6)_____ in the last decade.
	E	Mentions various works published in conference proceedings that extend DFMA principles in different ways.
	F	7)_____ that DFMA methods have mainly been applied to simple products and sub-assemblies, 8)_____ the analysis of assembly solutions and the goal of reducing the number of 9)_____.
Conclusion	G	Discusses the challenges in DFMA analysis when assembly technology 10)_____, which can impact the overall improvement.

Rise of a Great Power

Translate the following Chinese part into English and make it a complete English text.

1)(机械工程作为一门关键的工程学科,对于国家的制造业和技术发展具有重要意义)_____

_____.

As one of the world's largest manufacturing nations, China has achieved remarkable progress and accomplishments in the field of mechanical engineering.

From traditional mechanical manufacturing to CNC technology and automated production lines, 2)(中国的制造业正不断向智能化、高效化发展)_____

_____. Chinese mechanical engineers and technical teams have made significant contributions in areas such as mechanical design, process optimization, and manufacturing process improvement. 3)(例如,中国制造的高速列车、大型

工程机械和高精度机床等产品在国际上享有很高的声誉) _____
_____
_____ .

Moreover, China is committed to promoting sustainable development in mechanical engineering research. 4)(在能源和环境方面,中国的机械工程师正在开发和应用新的技术,以降低能源消耗和环境污染) _____
_____
_____ . Significantly, China has made notable progress in research and application of renewable energy sources such as solar and wind energy.

In conclusion, China's development in the field of mechanical engineering demonstrates strong capabilities and potential. 5)(未来,中国将继续加大在机械工程领域的投入,不断提升技术水平,推动机械工程行业的发展和创新) _____
_____
_____ .

1-1　Unit One-Answer and Translation

Unit Two Ecomechatronics

Warm-up

Implementing "Made in China 2025" to build a world manufacturing power is a vital strategic choice for us to learn from the historical lessons of missing the first two industrial revolutions and actively respond to the new round of technological revolution and industrial transformation. A new round of global scientific and technological revolution and industrial transformation brewing new breakthroughs, especially the deep integration of a new generation of information technology and manufacturing, coupled with breakthroughs in new energy, new materials, biotechnology, etc., is triggering far-reaching industrial transformation.

Thought-provoking Questions

· How can the integration of new-generation information technology and manufacturing processes drive the industrial transformation envisioned in "Made in China 2025"? What are the potential challenges and opportunities for mechanical engineering students in this context?

· How can mechanical engineering students contribute to the development and adoption of breakthrough technologies like new energy, new materials, and biotechnology, which are essential for driving the industrial transformation outlined in "Made in China 2025"?

Text A
Advancing Sustainable Machine Design

Research Background

Ecomechatronics is an engineering approach that focuses on developing and applying mechatronical technology to reduce the ecological impact and total cost of ownership of machines. It builds upon the integrative approach of mechatronics but with the additional goal of improving resource efficiency and minimizing environmental impact. The three key areas of machine improvement in ecomechatronics are energy efficiency, performance, and user comfort. This approach is driven by increasing awareness of resource scarcity and the need for sustainable development in manufacturing industries. Designing machines that use resources economically and provide higher energy efficiency and user comfort has become a market demand. Ecomechatronics requires an integrated design approach that considers the trade-off between energy efficiency, performance, and noise & vibrations. Key enabling technologies include energy-efficient components, design methods and tools, and advanced machine control techniques.

Introduction Questions

· What is the goal of ecomechatronics in machine development?

· What are the three key areas targeted for machine improvement in ecomechatronics?

· Why is there a growing demand for high-performance machines with improved energy efficiency and user comfort?

A. Ecomechatronics is an engineering approach to developing and applying mechatronical technology in order to reduce the ecological impact and total cost of ownership of machines. It builds upon the integrative approach of mechatronics, but not with the aim of only improving the functionality of a machine. Mechatronics is the multidisciplinary field of science and engineering that merges mechanics, electronics, control theory, and computer science to improve and optimize product design and manufacturing. In ecomechatronics, additionally, functionality should go hand in hand with an efficient use and limited impact on resources. Machine improvements are targeted in 3 key areas: energy efficiency, performance and user comfort (noise & vibrations).

B. Among policy makers and manufacturing industries there is a growing awareness of the scarcity of resources and the need for sustainable development. This results in new regulations with respect to the design of machines (e. g. European Ecodesign Directive 2009/125/EC) and to a paradigm shift in the global machines market: "instead of maximum profit from minimum capital, maximum added value must be generated from minimal resources". Manufacturing industries increasingly require high performance machines that use resources (energy, consumables) economically in a human-centered production. Machine building companies and original equipment manufacturers are thus urged to respond to this market demand with a new generation of high performance machines with higher energy efficiency and user comfort.

C. A reduction of the energy consumption lowers energy costs and reduces environmental impact. Typically more than 80% of the total-life-cycle impact of a machine is attributed to its energy consumption during the use phase. Therefore, improving a machine's energy efficiency is the most effective way of reducing its environmental impact. Performance quantifies how well a machine executes its function and is typically related to productivity, precision and availability. User comfort is related to the exposure of operators and the environment to noise & vibrations due to machine operation.

D. Since energy efficiency, performance and noise & vibrations are coupled in a machine they need to be addressed in an integrated way in the design phase. Example of the interrelation between the 3 key areas: with increasing machine speed typically the machine's productivity increases, but energy consumption will increase as well and machine vibrations may grow such that machine accuracy (e. g. positioning accuracy) and availability (due to downtime and maintenance) decrease. Ecomechatronical design deals with the trade-off between these key areas.

E. Ecomechatronics impacts the way mechatronical systems and machines are being designed and implemented. Therefore, the transformation to a new generation of machines concerns knowledge institutes, original equipment manufacturers, CAE software suppliers, machine builders and industrial machine owners. The fact that about 80% of the environmental impact of a machine is determined by its design puts emphasis on making the right technological design choices. A model-based, multidisciplinary design approach is required in order to address the energy efficiency, performance and user comfort of a machine in an integrated way.

F. The key enabling technologies can be categorized in machine components, machine design methods & tools, and machine control.

A few examples are listed below per category.

Machine components

• Energy efficient electrical motors: cf. energy efficiency classes of electric motors, ecodesign requirements for electric motors

• Variable frequency drives: variable motor speed enables energy reduction with respect to fixed speed applications

• Variable hydraulic pumps: energy reduction by adapting to required pressure and flow (e. g. variable displacement pump, load sensing pump)

• Energy storage technologies: electrical (battery, capacitor, supercapacitor), hydraulical (accumulator), kinetic energy (flywheel), pneumatic, magnetic (superconducting magnetic energy storage)

Design methods & tools

• Energetic simulations: using energetic machine models and empirical data (e. g. energy efficiency maps) to estimate the machine's energy consumption in the design phase

• Energy demand optimization: e. g. load leveling in order to avoid peaks in power demand

• Hybridization: applying at least one other, intermediate energy form in order to reduce primary power source consumption, e. g. in vehicles with internal combustion engines

• Vibro-acoustic analysis: study of the noise & vibrations signature of a machine in order to localize and differentiate between their root causes

• Multibody modeling: simulation of the interaction forces and displacements of coupled rigid bodies, e. g. to assess the effect of vibration dampers on a mechanical structure

• Active vibration damping: e. g. use of piezoelectric bearings for active control of machine vibrations

• Rapid control prototyping: provides a fast and inexpensive way for control and signal processing engineers to verify designs early and evaluate design tradeoffs

Machine control

• Energy consumption minimization: control signals are optimized for minimum energy consumption

· Energy management of energy storage systems: controlling the power flows and state-of-charge of an energy storage system with the aim of achieving maximum energy benefit and maximum system lifespan

· Model-based control: taking advantage of system models to improve the outcome (accuracy, reaction time, ...) of the controlled system

· (Self-)learning control: control self-adapting to the system and its changing environment, reducing the need for control parameter tuning and adaptation by the control engineer

· Optimal machine control: the control of the system is regarded as an optimization problem to which the control rules are considered the optimal solution.

Vocabulary List

multidisciplinary

involving or combining several academic disciplines or professional specializations 多学科的,跨学科的

vibrations

rapid back-and-forth movements or oscillations of an object or medium 震动,振动

downtime

the period of time during which a machine, system, or service is not functioning or unavailable 停工时间,停机时间

ecodesign

the practice of designing products with consideration for their environmental impact and sustainability 生态设计

hydraulic pumps

devices that convert mechanical energy into hydraulic energy, used to generate fluid flow or pressure 液压泵

accumulator

a device that stores potential energy in the form of a compressed gas or fluid, which can be released for use when needed 蓄能器,储能器

pneumatic

relating to or powered by compressed air or other gases 气动的,气压的

hybridization

the combination or integration of two or more different technologies, systems, or components 混合,混合化

damper

a device that reduces or absorbs vibrations, shocks, or fluctuations 减振器,阻尼器

parameter

a quantity or factor that can be measured or evaluated and is used as a basis for computation or decision-making 参数,参量

excavator

a heavy construction machine with a bucket or shovel at the front used for digging or moving large amounts of soil, rocks, or debris 挖掘机

supercapacitor

a type of capacitor that can store and release larger amounts of electrical energy than conventional capacitors 超级电容器

Language Enhancement

Ⅰ. Complete the following sentences with words listed in the box below. Change the form where necessary.

localize	verify	implement	simulation	assess
differentiate	interaction	estimate	categorize	maintenance
intermediate	evaluate	commercialize	integrate	generate

1. It is essential to _____ the ecological impact and total cost of ownership of machines in ecomechatronics.

2. Mechatronics _____ mechanics, electronics, control theory, and computer science to optimize product design and manufacturing.

3. The authenticity of the data needs to be _____ before drawing any conclusions.

4. The company plans to _____ their innovative product by next year.

5. The engineers ran multiple _____ to optimize the design of the bridge for maximum stability.

6. The technician will _____ the mechanical components of the machine to determine the cause of the malfunction.

7. Engineers use specialized tools to _____ between various electronic components in order to troubleshoot and repair circuitry.

8. In mechatronics, the _____ between mechanical systems and electrical components is crucial for the proper functioning of robotic arms.

9. Engineers often _____ the lifespan of mechatronic devices by analyzing the wear and tear on their moving parts.

10. To simplify troubleshooting, engineers _____ mechatronic systems into subgroups based on their primary functions, such as sensing, actuation, or control.

11. Regular _____ of mechatronic systems is essential to ensure their long-term reliability and performance.

12. The _____ stages of a mechatronics project involve integrating mechanical, electrical, and software components to create a functional prototype.

13. Engineers use advanced sensors to _____ the position of a mechatronic system within its environment, enabling precise navigation and interaction.

14. After thorough testing and analysis, engineers _____ the optimized control algorithms into the mechatronic system to enhance its overall performance.

15. The mechatronic system is designed to _____ real-time feedback and data, allowing engineers to monitor and analyze its operation for continuous improvement.

Ⅱ. **Complete the following sentences with phrases listed in the box below. Change the form where necessary.**

be coupled with	adapt to	be determined by	take advantage of
with respect to	attribute to	put emphasis on	

1. The success of a software project can _____ the effectiveness of its project management.

2. The company _____ innovation and cutting-edge technology in its product development.

3. We should _____ artificial intelligence to automate repetitive tasks and improve efficiency.

4. _____ cybersecurity, it is crucial to implement strong encryption measures to protect sensitive data.

5. The performance of a computer system _____ its hardware specifications and software optimization.

6. In order to stay competitive, businesses need to _____ the rapid advancements in digital technology.

7. The increase in productivity can _____ the implementation of a new software system.

Ⅲ. Complete the summary with words listed in the box below. Change the form where necessary.

impact	simulation	shift	sustainability	collaboration
demonstrate	integrate	holistic	interrelation	consume

Ecomechatronics reduces ecological impact and ownership costs. It 1)_____ mechatronics principles, focusing on energy efficiency, performance, and user comfort. Global awareness of 2)_____ has led to new regulations and market 3)_____ towards resource-efficient machines. Energy 4)_____ contributes over 80% to a machine's life-cycle 5)_____, making energy efficiency crucial. Integrated design addresses the 6)_____ between energy efficiency, performance, and noise/vibrations. 7)_____ among institutes, manufacturers, software suppliers, builders, and owners is vital. Key enabling technologies include efficient motors, variable drives, 8)_____, analysis, and control. Applications like the Komatsu PC200-8 Hybrid excavator 9)_____ ecomechatronics in action. By integrating energy efficiency, performance, and comfort, ecomechatronics offers a 10)_____ approach to sustainable machine design.

Academic Expression

Pair work: Discuss with your partner and compare the two possible paraphrases of each sentence and decide which one expresses the original meaning more academically.

1. Ecomechatronics is an engineering approach that aims to reduce the ecological impact and total cost of ownership of machines by developing and applyingmechatronical technology.

    a. Ecomechatronics is an engineering methodology employed to mitigate the ecological footprint and overall cost of machine ownership through the development and application of mechatronical technology.

    b. Ecomechatronics is an engineering approach that focuses on reducing the environmental impact and overall cost of owning machines. It achieves this by developing and utilizing mechatronical technology.

2. Mechatronics merges mechanics, electronics, control theory, and computer science to optimize product design and manufacturing.

 a. Mechatronics combines mechanics, electronics, control theory, and computer science to improve product design and manufacturing.

 b. Mechatronics is an interdisciplinary field integrating mechanics, electronics, control theory, and computer science to enhance the efficiency of product design and manufacturing processes.

3. Ecomechatronics targets energy efficiency, performance, and user comfort (noise &vibrations) as key areas for machine improvements.

 a. Ecomechatronics focuses on enhancing energy efficiency, performance, and user comfort (regarding noise and vibrations) as crucial aspects for advancing machine capabilities.

 b. In ecomechatronics, the primary areas of focus for improving machines are energy efficiency, performance, and ensuring user comfort by minimizing noise and vibrations.

4. There is a growing awareness among policy makers and manufacturing industries about the need for sustainable development.

 a. Policy makers and manufacturing industries are becoming more aware of the importance of sustainable development.

 b. Policy makers and manufacturing industries are increasingly recognizing the imperative of sustainable development.

5. The reduction of energy consumption is an effective way to minimize a machine's environmental impact.

 a. Reducing energy consumption constitutes a viable approach to mitigate the environmental repercussions of machines.

 b. Cutting down on energy consumption is an effective way to reduce the environmental impact of machines.

6. The interrelation between energy efficiency, performance, and noise & vibrations in machines necessitates an integrated approach during the design phase.

a. Because energy efficiency, performance, and noise/vibrations are interconnected in machines, they need to be addressed together during the design phase.

b. The interdependence of energy efficiency, performance, and noise and vibrations in machines underscores the requirement for an integrated design approach.

7. The transformation to a new generation of machines necessitates collaboration among various stakeholders.

a. The transition to a new era of machines mandates collaboration among diverse stakeholders.

b. Shifting to a new generation of machines requires cooperation among different stakeholders.

8. Key enabling technologies in ecomechatronics include machine components, design methods and tools, and machine control.

a. Critical technologies in ecomechatronics cover machine components, design methods and tools, and machine control.

b. Essential technologies in ecomechatronics encompass machine components, design methodologies and tools, as well as machine control systems.

Understanding the Text

Ⅰ. **Pair work: Work with your partner and take turns asking and answering the following questions according to the information contained in the text.**

1. What is the concept of ecomechatronics?

2. What are the three key areas targeted for improvement in ecomechatronics?

3. Why is improving a machine's energy efficiency important in reducing its environmental impact?

4. How are energy efficiency, performance, and noise/vibrations related in a machine?

5. What are some key enabling technologies in ecomechatronics?

6. What is the importance of making the right technological design choices in machine design?

_____

7. How does ecomechatronics impact various stakeholders in the machines industry?

_____

8. What are some examples of ecomechatronical system applications?

_____

II. Identify the paragraph from which the information is derived. You may use any letter more than once.

1. Ecomechatronics influences the design and implementation of mechatronical systems and machines, emphasizing the importance of making appropriate design choices.

2. Lowering energy consumption reduces costs and environmental impact, with over 80% of a machine's life-cycle impact attributed to energy use during operation.

3. Machine components in ecomechatronics include energy-efficient electrical motors, variable frequency drives, variable hydraulic pumps, and various energy storage technologies.

4. Key enabling technologies in ecomechatronics include machine components, design methods & tools, and machine control.

5. Design methods & tools in ecomechatronics involve energetic simulations, energy demand optimization, hybridization, vibro-acoustic analysis, multibody modeling, active vibration damping, and rapid control prototyping.

6. Ecomechatronics aims to reduce the ecological impact and ownership cost of machines through mechatronical technology. It goes beyond improving functionality.

7. Ecomechatronical systems find applications in various areas, such as hybrid excavators, hybrid buses, and hybrid tram vehicles.

8. Machine control in ecomechatronics focuses on minimizing energy consumption, managing energy storage systems, utilizing models, implementing self-learning control, and optimizing control rules.

9. Policy makers and industries recognize the scarcity of resources and the need for sustainable development, leading to new regulations and a shift towards maximum added value with minimal resources.

10. Energy efficiency, performance, and noise & vibrations are interconnected in machine design, requiring an integrated approach.

Rise of a Great Power

Translate the following Chinese part into English and make it a complete English text.

China has made significant progress in advancing sustainable machine design. The government has established energy-saving policies and standards, encouraging enterprises to focus on energy efficiency. Companies have adopted efficient motors and optimized energy management to reduce energy consumption, 1)（此举不仅降低成本，还减少碳排放和环境污染）_____ _____ _____.

In terms of sustainable material selection, China's manufacturing industry has begun replacing traditional materials with environmentally friendly alternatives, such as renewable materials and biodegradable materials. This reduces reliance on natural resources and decreases environmental impact. Intelligent control technology represents another breakthrough in China. 2)（引入人工智能和物联网技术，使机器设计更智能化、自动化）_____ _____. Intelligent control systems monitor machine status in real-time and make adaptive adjustments to maximize efficiency and resource utilization. 3)（智能化机器设计减少能源和材料浪费，提高生产效率和产品质量） _____ _____.

4)（中国在可持续机器设计方面的进展为环境保护和可持续发展提供了宝贵经验）_____ _____. By focusing on energy efficiency, material selection, and intelligent control technology, 5)（中国制造业正朝着更环保、高效和可持续的方向发展）_____ _____. This will contribute significantly to the construction of a more sustainable future.

Text B
Supervised Learning: ML Model Development Basics

Research Background

Supervised learning is a widely implemented machine learning (ML) method where ML models learn functions by fitting inputs to outputs. It involves training the models on categorized training data to uncover patterns and make novel observations. Linear regression is a commonly used predictive statistical analysis in supervised learning, and the choice of algorithm depends on the linearity or nonlinearity of the problem. The process includes steps such as dataset acquisition and processing, target variable selection, dataset splitting, and hyper-parameter tuning for better predictions.

Introduction Questions

· What is the main objective of supervised learning in machine learning?
· Why is dataset acquisition and processing important in ML?
· What is the purpose of splitting the dataset in supervised learning?

A. Supervised learning is the most commonly implemented ML method. In this method, ML models need to learn functions in a way that inputs fit the outputs. Then, the function reveals information from categorized training data and each input is related to its assigned value. The algorithm embedded in a ML model is capable of making novel observations never made before or uncovering patterns in a training data set.

B. One of the most approached predictive statistical analysis in supervised method is linear regression. Each analysis is carried out by a specific algorithm which should be chosen with prior knowledge based on linearity or nonlinearity of the problem. However, by comparing error metrics of each regression, best algorithm could be identified.

In the following, the aforementioned steps are discussed in more detail.

Dataset acquisition and processing

C. Data acquisition can be regarded as a concept where physical events that happened in real world gets transformed into electrical signals, converted, and scaled in digital format for further analysis, processing, and storage within the computer memory storage. In general data acquisition systems are not only for gathering data but also for operating on the data. Having complete data is very important for ML models to perform better and get a robust analysis. Nowadays, DL models can even operate as good as real ophthalmologists in detecting diabetic eye issues from an image, all owning to the computational power of models and large amount of data to train the models.

D. With modernization and new fields of science used in the industry, lack of prior data is a problem that should be dealt with, particularly with deep learning models that require even more data than traditional ML. Initially, data acquisition approaches are used to harvest, augment, or generate of novel data sets. Afterwards, data labeling should be done and then, training the already achieved or improve the labeling and accuracy of the gathered dataset. In this aspect, ML engineers and data scientists and data managers should work together. In the following, a diagram of steps in data collection, acquisition and processing is presented.

Target variable selection

E. As the name indicates, the target variable is the feature that we aim to get or achieve in the ML task. Whether classification or regression, the features should be clear, such as the target variable. Target variables in the form of labeled targets are the pivotal point where supervised ML algorithms use historical data to pick apart patterns and discover relations amongst the other unknown features of the dataset and the set target variable. Without properly labeled data, supervised ML tasks would not be able to plot data to outcomes.

Splitting the dataset

F. Based on most references and as a convention, it is understood that it is best to split the dataset to prevent overestimation and overfitting. In the following, we discuss the most noticeable sets of grouping: training set, cross-validation set, and testing test. The training test is mostly the largest set. The model trains based on the insight that it gains from the data that is fed from the training set. The training set is basically the subset of the whole data set available. In this phase, we can forecast the weights, the bias of the model if It's a NN.

G. Therefore, we can optimize the hyper-parameters, which are the parameters that control the initial setting of the system. They are immensely important because after setting them, they cannot be changed like the weights or biases or the parameters of the system. In the cross-validation phase, we estimated the loss function or error of the system, and therefore minimizing it to get the best prediction. And then finally, we use the testing set, which is the smallest of the aforementioned sets and results in a non-bias result because the testing data are new to the model. This stage acts as a close simulation to a real life occurrence and demonstrates how the model would operate in a real situation.

Hyper-parameter tuning and prediction

H. Hyper-parameters are very important due to the fact that they should be set before each iteration, and they define the very fundamentals of an ML model, unlike process parameters that can be manipulated while data learning process is in process. In the case of a DNN, a part of it is determining the number of hidden layers, nods, neurons, step size, and batch size. One needs to differentiate between the hyper-parameters that are related to the algorithm, such as the aforementioned step size, batch size, and the ones that are related to the structure of the model, such as the number of hidden layers, method of nods connecting to each other and the number of nods. As is maintained, hyper-parameters are constant while in operation, but process parameters can change.

I. The progression to tuning or optimization of hyper-parameters could be achieved when enough number of tests runs and trials are undertaken. The pace of training of a DL model is determined by the rate of convergence. There are methods known as super convergence methods where the crucial foundations of super-convergence are training with a singular learning rate cycle and a hefty maximum learning rate.

J. By comparison of the results of the test runs and making vigilant comparisons to real values of each data iteration, the accuracy of the model can be evaluated, and thereafter, we gain insight as to find the best values for the system to make a better combination of hyper-parameters and more accurate predictions. A hyper-parameter metric is a personal specification of a single target variable that is specified by choice of a human operator. The model accuracy is defined by a metric value and, therefore, can be determined if it is maximization or minimization that is the desired goal for our specified metric to fulfill.

Vocabulary List

algorithm

a set of rules or procedures for solving a problem　算法

embed

to incorporate or integrate something into another thing　嵌入；内嵌

linear

relating to or resembling a straight line; involving one variable　线性的

regression

a statistical method for modeling the relationship between variables　回归

nonlinearity

the quality or state of not being linear　非线性性质

ophthalmologist

a medical doctor specializing in eye care　眼科医生

dataset

a collection or set of data　数据集

overfitting

the excessive fitting or modeling of data to a particular function　过度拟合

cross-validation set

a subset of data used for validating a model's performance　交叉验证集

hyper-parameters

parameters that define the structure or behavior of a model　超参数

Language Enhancement

I. **Complete the following sentences with words listed in the box below. Change the form where necessary.**

identify	supervise	harvest	manipulate	convert
immense	optimize	embed	differentiate	augment
format	fulfill	transform	vigilant	demonstrate

1. Machine learning algorithms can _____ the task of classifying images with high accuracy.

2. It is important to be _____ in monitoring the performance of machine learning models to detect any anomalies or biases.

3. One key aspect of machine learning is the ability to _____ between different types of data, such as numerical and categorical variables.

4. Feature manipulation is a crucial step in machine learning, as it involves transforming and _____ data to extract meaningful patterns.

5. The researcher used a dataset to _____ the effectiveness of the machine learning algorithm in predicting customer behavior.

6. The amount of data available for training machine learning models can be _____, requiring efficient storage and processing resources.

7. Model _____ techniques aim to enhance the performance of machine learning algorithms by fine-tuning their parameters.

8. _____ techniques are used to represent textual data in a numerical format suitable for machine learning algorithms.

9. Machine learning can be applied to _____ insights from large amounts of unstructured data, such as social media posts or customer reviews.

10. By incorporating additional features, it is possible to _____ the predictive power of a machine learning model.

11. Data _____ plays a crucial role in preparing input data for machine learning algorithms, ensuring compatibility and consistency.

12. Machine learning algorithms can _____ patterns and trends in data that may not be readily apparent to human analysts.

13. Through the process of _____, raw data can be converted into a format suitable for training machine learning models.

14. It is essential to have proper _____ and monitoring in place to ensure the ethical and responsible use of machine learning algorithms.

15. Machine learning models can _____ input data into predictions or recommendations, enabling automated decision-making processes.

Ⅱ. Complete the following sentences with phrases listed in the box below. Change the form where necessary.

transform into	lack of	be determined by	be specified by	owe to
be fed from	carry out	regard as	deal with	be related to

1. The research findings show that the genetic mutation _____ the development of certain diseases.

2. The experiment will _____ in a controlled laboratory environment to ensure accurate results.

3. Artificial intelligence _____ a game-changing technology with vast potential.

4. The company plans to _____ their traditional manufacturing processes _____ automated systems to improve efficiency.

5. The success of the project _____ the dedication and hard work of the team members.

6. _____ funding has hindered the progress of scientific research in this field.

7. The engineers are working tirelessly _____ the complex technical challenges that arise during the development of the new software.

8. The data used for analysis in this study will _____ various sensors placed throughout the testing site.

9. The outcome of the experiment will _____ the variables and parameters set in the research design.

10. The safety protocols for the chemical laboratory _____ the international standards to ensure a secure working environment.

Ⅲ. Complete the summary with words listed in the box below. Change the form where necessary.

prediction	recommend	split	augmentation	determination
absence	encompass	acquisition	guide	categorize
regression	regression	convergence	supervise	evaluate

Supervised learning is a common machine learning method where ML models learn functions that map inputs to outputs based on 1) _____ training data. Linear 2) _____ is a widely used algorithm for supervised learning, chosen based on problem linearity. Data 3)_____ involves transforming real-world events into digital format for analysis. Sufficient and robust data is essential for ML models, and in the 4)_____ of prior data, approaches like data 5)_____ or generation are employed. Labeled target variables play a key role in 6)_____ ML, enabling pattern identification and relationship discovery. Dataset splitting is 7)_____ to avoid overfitting, creating training, cross-validation, and testing sets. The training set trains the model, the cross-validation set 8)_____ predictions, and the testing set 9)_____ model performance on new data. Hyper-parameter tuning, crucial for model development, involves selecting parameters related to algorithm and model structure. Super 10)_____ methods, employing specific learning rate cycles, aid training pace 11)_____. Accuracy evaluation involves comparing test results with real values and optimizing hyper-parameters for improved 12)_____. Metrics define accuracy and 13)_____ the desired goal of maximization or minimization. This summary outlines the steps of supervised learning, 14)_____ data acquisition, target variable selection, dataset 15)_____, hyper-parameter tuning, and prediction evaluation.

Academic Expression

Pair work: Discuss with your partner and compare the two possible paraphrases of each sentence and decide which one expresses the original meaning more academically.

1. Supervised learning is commonly used in machine learning (ML), where ML models learn to map inputs to outputs based on categorized training data.

  a. In machine learning (ML), supervised learning is a widely used technique where ML models learn to connect inputs with outputs using categorized training data.

  b. Supervised learning is a widely utilized technique in machine learning (ML), where ML models are trained to establish mappings between inputs and outputs based on categorized training data.

2. Linear regression is a popular predictive statistical analysis method in supervised learning, and the choice of the algorithm depends on the linearity or nonlinearity of the problem.

  a. Linear regression is a prevalent method for predictive statistical analysis in supervised learning. The selection of the algorithm for each analysis depends on the linearity or nonlinearity of the problem at hand.

  b. Linear regression is a popular method for making predictions in supervised learning. The choice of the algorithm depends on whether the problem is more linear or nonlinear.

3. Data acquisition involves converting real-world events into electrical signals, converting and scaling them into a digital format for analysis, processing, and storage.
 a. Data acquisition involves converting real-world events into electrical signals, which are then transformed, scaled, and converted into a digital format for analysis, processing, and storage.
 b. Data acquisition involves the conversion of real-world events into electrical signals, followed by their conversion, scaling, and digital formatting for subsequent analysis, processing, and storage.

4. The target variable in ML tasks represents the feature we aim to achieve, and it is crucial for supervised ML algorithms to identify patterns and relations in the dataset.
 a. The target variable in ML tasks represents the specific feature that is aimed to be achieved. It plays a crucial role in supervised ML algorithms for the identification of patterns and relationships within the dataset.
 b. In ML tasks, the target variable represents the specific feature that we want to achieve. It's really important for ML algorithms to find patterns and relationships in the dataset.

5. It is recommended to split the dataset into training, cross-validation, and testing sets to prevent overestimation and overfitting in ML models.
 a. To avoid making overly optimistic predictions and fitting the data too closely, It's a good idea to divide the dataset into training, cross-validation, and testing sets.
 b. To mitigate the risks of overestimation and overfitting in ML models, it is recommended to partition the dataset into distinct subsets, namely training, cross-validation, and testing sets.

6. Hyper-parameters play a crucial role in ML models as they define the fundamental settings of the system and need to be optimized to achieve better predictions.
 a. Hyper-parameters hold significant importance in ML models, as they define the essential configurations of the system and necessitate optimization to enhance predictive performance.
 b. Hyper-parameters are super important in ML models because they define the fundamental settings of the system. We need to find the best values for these parameters to improve predictions.

7. Tuning hyper-parameters involves selecting the optimal values for parameters that control the ML model's behavior and structure.

    a. Tuning hyper-parameters means choosing the best values for the parameters that control how the ML model behaves and is structured.

    b. The process of tuning hyper-parameters involves the careful selection of optimal parameter values that govern the behavior and structure of the ML model.

8. Evaluating the accuracy of the ML model involves comparing the results of test runs to real values, allowing us to determine the best combination of hyper-parameters and improve predictions.

    a. The evaluation of ML model accuracy encompasses the comparison of test results with real values, enabling the determination of the most effective combination of hyper-parameters to refine predictions.

    b. To evaluate how accurate an ML model is, we compare the results of test runs to real values. This helps us figure out the best combination of hyper-parameters to make more accurate predictions.

Understanding the Text

Ⅰ. Identify the paragraph from which the information is derived. You may choose a paragraph more than once.

1. Various approaches are employed for data acquisition, including the collection, augmentation, and generation of novel datasets to address the shortage of prior data.

2. Hyper-parameters, including settings related to the algorithm and the model's structure, must be carefully tuned and optimized before the model's iteration to achieve the best possible results.

3. Properly defining and labeling the target variable is crucial for effective supervised machine learning algorithms to identify patterns and establish correlations within the dataset.

4. Hyper-parameters hold significant importance as they define the foundational aspects of an ML model and remain constant.

5. Supervised learning stands out as the most commonly implemented method in the field of machine learning, finding extensive application in various domains.

6. To prevent overestimation and overfitting, it is a common practice to split the dataset into distinct subsets, such as the training set, cross-validation set, and testing set.

7. To evaluate the model's performance, it is essential to compare the outcomes of test runs with actual values and analyze the results using appropriate metrics.

8. The process of acquiring and processing datasets involves converting real-world events into electrical signals, which are subsequently transformed and scaled into digital format.

9. In the ever-evolving landscape of modernization and diverse scientific fields, one of the challenges faced in the industry is the scarcity of prior data.

10. Optimizing hyper-parameters requires a systematic process involving multiple test runs and trials to identify the optimal configuration.

Ⅱ. **In this section, there are eight statements. Do these statements agree with the information given in Text B? You should decide on the best choice.**

TRUE   if the statement agrees with the information
FALSE   if the statement contradicts the information
NOT GIVEN   if there is no information on this

1. Supervised learning is the most commonly implemented ML method.
2. Linear regression is a commonly used algorithm.
3. Data acquisition systems are not only used for gathering data but also for operating on the data.
4. Acquiring labeled data is crucial for supervised ML tasks to analyze patterns and relations.
5. Deep learning models require even less data than traditional ML models.
6. Splitting the dataset is important to prevent overestimation and overfitting.
7. Hyper-parameters can be adjusted during the data learning process.
8. The rate of convergence determines the pace of training of learning model.

Ⅲ. **Group work: Please work in groups and complete the structural table with the information from Text B.**

Section	Paragraph	Content
Introduction	A	Introduces 1)_____ learning and its use in machine learning. Highlights the function of ML models in learning from 2)_____ training data.
	B	Discusses linear 3)_____ as a common predictive statistical analysis method in supervised learning. Emphasizes the choice of 4)_____ based on linearity or nonlinearity.
Body	C	Explains data 5)_____, where real-world events are converted into electrical signals and processed digitally.
	D	Addresses the 6)_____ of limited prior data and the use of data acquisition approaches to generate or 7)_____ datasets.
	E	Highlights the importance of the target variable in machine learning tasks and its role in 8)_____ patterns and relationships.
	F	Discusses the 9)_____ of splitting the dataset into training, cross-validation, and testing sets to prevent 10)_____ and overfitting.
	G	Emphasizes the 11)_____ of hyper-parameters and the use of cross-validation for accurate predictions.
Conclusion	H	Describes the significance of hyper-parameters in 12)_____ ML models and their constant nature during operation.
	I	Discusses the process of tuning hyper-parameters through testing and the concept of super 13)_____.
	J	Explores the 14)_____ of model accuracy and the importance of finding 15)_____ system values for improved predictions.

Rise of a Great Power

Translate the following Chinese part into English and make it a complete English text.

In recent years, China has made significant progress in the development of machine learning models. 1)(许多中国的科研机构和高校在机器学习领域进行前沿的研究,提出了许多创新的算法和模型)_____

_____. These algorithms and models have achieved remarkable results in areas such as speech recognition, image processing, and natural language processing.

Numerous universities and research institutions have established specialized AI and machine learning institutes, offering relevant degrees and training courses. Additionally, 2)(许多企业也开展了人才培养计划,吸引了大量优秀的人工智能专业人才)_____

_____.

Furthermore, 3)(中国在机器学习应用领域的探索中也取得了重要的进展,比如在金融、医疗、交通等方面取得了显著的效果)_____

_____. These applications have not only enhanced work efficiency but also created additional business value for companies.

The development of machine learning technology in China is driven by the support of the government, the efforts of research institutions, and the innovation of enterprises. This has laid a solid foundation for China's rise in the field of artificial intelligence. Looking ahead, we can expect 4)(中国在机器学习领域取得更多的成就和突破,为推动全球的科技创新和发展做出更大的贡献)_____

_____.

2-1　Unit Two-Answer and Translation

Unit Three　Biomechatronics

Warm-up

Many developing countries are also speeding up planning and layout, actively participating in the global industrialredivision of labor, and seeking a favorable position for a new round of competition. These countries understand that engaging in industrial activities can lead to economic growth, job creation, and technological advancements. By positioning themselves strategically, they can attract foreign investments, establish partnerships with multinational corporations, and gain access to new markets. This proactive approach enables developing countries to harness their potential and contribute to the global economy's dynamism and diversification.

Thought-provoking Questions

· As a mechanical engineering student, how can you contribute to the industrial development of your country and help position it strategically in the global competition? What specific skills or knowledge areas would be crucial in attracting foreign investments and fostering technological advancements?

· In the context of global industrial redivision of labor, how can developing countries strike a balance between economic growth, job creation, and environmental sustainability? As a mechanical engineering student, what approaches or technologies can you propose to address this challenge and promote sustainable industrial development?

Text A
Biomechatronics: Bridging Biology and Engineering

Research Background

Biomechatronics is an interdisciplinary field that integrates biology, mechatronics, robotics, and neuroscience. It encompasses a wide range of applications, from prosthetic limbs to engineering solutions for respiration, vision, and the cardiovascular system. The field works by mimicking how the human body functions, using biosensors to detect user intentions and motions, electromechanical sensors to measure device information, controllers to relay user intentions and interpret feedback, and actuators to produce force and movement. Biomechatronics research focuses on analyzing human motions, interfacing electronic devices with the nervous system, and using living muscle tissue as actuators. Neural interfacing allows devices to connect with the user's muscles and nerves, enabling advanced control. Despite the growing demand, biomechatronic technologies face challenges such as high costs, limited insurance coverage, mechanical reliability, and neural connections.

Introduction Questions

· What is the purpose of the controller in a biomechatronic device?
· What is the significance of neural interfacing in biomechatronics?
· What challenges do biomechatronic technologies face in the healthcare market?

Biomechatronics is an applied interdisciplinary science that aims to integrate biology and mechatronics (electrical, electronics, and mechanical engineering). It also encompasses the fields of robotics and neuroscience. Biomechatronic devices encompass a wide range of applications from the development of prosthetic limbs to engineering solutions concerning respiration, vision, and the cardiovascular system.

A. How it works

Biomechatronics mimics how the human body works. For example, four different steps must occur to be able to lift the foot to walk. First, impulses from the motor center of the brain are sent to the foot and leg muscles. Next, the nerve cells in the feet send information, providing feedback to the brain, enabling it to adjust the muscle groups or amount of force required to walk across the ground. Different amounts of force are applied depending on the type of surface being walked across. The leg's muscle spindle nerve cells then sense and send the position of the floor back up to the brain. Finally, when the foot is raised to step, signals are sent to muscles in the leg and foot to set it down.

B. Biosensors

Biosensors are used to detect what the user wants to do or their intentions and motions. In some devices the information can be relayed by the user's nervous system or muscle system. This information is related by the biosensor to a controller which can be located inside or outside the biomechatronic device. In addition, biosensors receive information about the limb position and force from the limb and actuator. Biosensors come in a variety of forms. They can be wires which detect electrical activity, needle electrodes implanted in muscles, and electrode arrays with nerves growing through them.

C. Electromechanical sensors

The purpose of the mechanical sensors is to measure information about the biomechatronic device and relate that information to the biosensor or controller. Additionally, many sensors are being used at schools, such as Case Western Reserve University, the University of Pittsburgh, Johns Hopkins University, among others, with the goal of recording physical stimuli and converting them to neural signals for a subarea of biomechatronics called neuromechatronics.

D. Controller

The controller in a biomechatronic device relays the user's intentions to the actuators. It also interprets feedback information to the user that comes from the biosensors and mechanical sensors. The other function of the controller is to control the biomechatronic device's movements.

E. Actuator

The actuator can be an artificial muscle but it can be any part of the system which provides an outward effect based on the control input. For a mechanical actuator, its job is to produce force and movement. Depending on whether the device is orthotic or prosthetic the actuator can be a motor that assists or replaces the user's original muscle. Many such systems actually involve multiple actuators.

F. Research

Biomechatronics is a rapidly growing field but as of now there are very few labs which conduct research. The Shirley Ryan AbilityLab (formerly the Rehabilitation Institute of Chicago), University of California at Berkeley, MIT, Stanford University, and University of Twente in the Netherlands are the researching leaders in biomechatronics. Three main areas are emphasized in the current research.

- Analyzing human motions, which are complex, to aid in the design of biomechatronic devices
- Studying how electronic devices can be interfaced with the nervous system.
- Testing the ways to use living muscle tissue as actuators for electronic devices

G. Neural Interfacing

Interfacing allows biomechatronics devices to connect with the muscle systems and nerves of the user in order to send and receive information from the device. This is a technology that is not available in ordinary orthotics and prosthetics devices. Groups at the University of Twente and University of Malaya are making drastic steps in this department. Scientists there have developed a device which will help to treat paralysis and stroke victims who are unable to control their foot while walking. The researchers are also nearing a breakthrough which would allow a person with an amputated leg to control their prosthetic leg through their stump muscles.

Researchers at MIT have developed a tool called the MYO-AMI system which allows for proprioceptive feedback (position sensing) in the lower extremity (legs, transtibial). Still others focus on interfacing for the upper extremity (Functional Neural Interface Lab, CWRU). There are both CNS and PNS approaches further subdivided into brain, spinal cord, dorsal root ganglion, spinal/cranial nerve, and end effector techniques and some purely surgical techniques with no device component.

H. Growth

The demand for biomechatronic devices are at an all-time high and show no signs of slowing down. With increasing technological advancement in recent years, biomechatronic researchers have been able to construct prosthetic limbs that are capable of replicating the functionality of human appendages. Such devices include the "i-limb", developed by prosthetic company Touch Bionics, the first fully functioning prosthetic hand with articulating joints, as well as Herr's Power Foot BiOM, the first prosthetic leg capable of simulating muscle and tendon processes within the human body. Biomechatronic research has also helped further research towards understanding human functions. Researchers from Carnegie Mellon and North Carolina State have created an exoskeleton that decreases the metabolic cost of walking by around 7 percent.

Despite the demand, however, biomechatronic technologies struggle within the healthcare market due to high costs and lack of implementation into insurance policies. Herr claims that Medicare and Medicaid specifically are important "market-breakers or market-makers for all these technologies," and that the technologies will not be available to everyone until the technologies get a breakthrough. Biomechatronic devices, although improved, also still face mechanical obstructions, suffering from inadequate battery power, consistent mechanical reliability, and neural connections between prosthetics and the human body.

Vocabulary List

biomechatronics
an applied interdisciplinary science integrating biology, mechatronics, robotics, and neuroscience 生物机电一体化

prosthetic limbs
artificial limbs designed to replace missing body parts 假肢

spindle nerve cells
specialized nerve cells that sense position and movement 肌梭神经细胞

biosensors
devices that detect and measure biological information 生物传感器

actuator
a mechanism that produces movement or action 作动器

neuromechatronics
the study of the integration of neuroscience and mechatronics 神经机电一体化

interfacing
the act of connecting or linking two systems or components together 接口

transtibial

relating to or involving the area below the knee and above the ankle　胫骨下的

ganglion

a mass of nerve tissue located outside the brain or spinal cord　神经节

appendage

a part or an organ that is attached to the main structure of an organism　附肢

tendon

a flexible cord of strong tissue that connects muscle to bone　腱

exoskeleton

a hard external skeleton that provides support and protection　外骨骼

electrode

a conductor used to make electrical contact with a nonmetallic part of a circuit　电极

orthotic

relating to the design and fitting of orthopedic devices　矫形的

rehabilitation

the process of restoring someone's health or abilities through therapy and training　康复

metabolic

relating to the chemical processes occurring within a living organism　新陈代谢的

Language Enhancement

Ⅰ. Complete the following sentences with words listed in the box below. Change the form where necessary.

encompass	impulse	multiple	implant	detect
respiration	rehabilitation	approach	involve	relay
original	artificial	convert	drastic	paralysis
available	subdivide	adjust	integrate	replicate

1. Scientists are trying to _____ the experiment to validate the results.
2. The landowner decided to _____ the large property into smaller lots for sale.
3. The _____ to solving this problem requires careful analysis and creative thinking.
4. The company had to make _____ cuts in order to reduce costs and stay afloat.
5. The accident left him with partial _____ in his legs.
6. There are limited seats _____ for the concert, so make sure to book yours early.
7. After the surgery, the patient underwent extensive _____ to regain mobility.
8. The project involves _____ teams working together to achieve the desired outcome.
9. The _____ race requires each team member to pass the baton to the next runner.
10. The painting's value lies in its _____ and unique brushstrokes.
11. _____ intelligence is revolutionizing various industries with its advanced capabilities.
12. We need to _____ the document from PDF to Word format for editing.
13. The surgeon _____ a pacemaker to regulate the patient's heartbeat.
14. The new technology can _____ even the smallest traces of contaminants in water.
15. _____ is the process of inhaling oxygen and exhaling carbon dioxide.
16. The nervous system transmits electrical _____ to coordinate bodily functions.
17. It took some time for her to _____ to the new school environment.
18. The goal is to _____ different software systems into a seamless workflow.
19. The project aims to _____ various aspects of sustainable development.
20. The development of self-driving cars _____ a complex network of sensors, algorithms, and machine learning techniques.

Ⅱ. **Complete the following sentences with phrases listed in the box below. Change the form where necessary.**

subdivide into	suffer from	occur to	base on
aim to	depend on	focus on	be interfaced with

1. Our research team _____ develop a breakthrough technology that revolutionizes renewable energy production.
2. It didn't _____ the scientists that the newly discovered element could have such significant implications for medical imaging.
3. The success of the software project _____ the efficiency of the underlying algorithms.

4. The design of the new smartphone _____ the latest advancements in mobile technology.

5. The smart home system _____ various devices, allowing users to control their home appliances remotely.

6. The company _____ developing cutting-edge artificial intelligence solutions for autonomous vehicles.

7. The research project _____ different phases to address specific technological challenges.

8. He _____ digital eye strain due to excessive screen time and improper use of electronic devices.

Ⅲ. Complete the summary using the list of words in the box. Change the form where necessary.

impulse	adjustment	integrate	analyze	connect
actuator	various	significant	promise	interface
mimic	detecting	accessible	interpret	relay

Biomechatronics is an interdisciplinary field that aims to 1) _____ biology, mechatronics, robotics, and neuroscience. It involves the development of biomechatronic devices for 2) _____ applications, including prosthetic limbs and engineering solutions related to respiration, vision, and the cardiovascular system.

The functioning of biomechatronic devices 3) _____ the workings of the human body. It involves steps such as motor 4) _____ from the brain, feedback from nerve cells, and muscle 5) _____ based on surface conditions. Biosensors play a crucial role in 6) _____ user intentions and motions, 7) _____ information to a controller. Mechanical sensors measure device information and 8) _____ with biosensors or controllers.

The controller 9) _____ user intentions and feedback information, while the actuator produces force and movement. Research in biomechatronics focuses on 10) _____ human motions, interfacing electronic devices with the nervous system, and using living muscle tissue as 11) _____. Notable research institutions include the Shirley Ryan AbilityLab, MIT, Stanford University, and the University of Twente.

Neural interfacing allows biomechatronic devices to 12) _____ with the user's muscles and nerves, enabling bidirectional information exchange. 13) _____ advancements include devices for paralysis and stroke victims, as well as control of prosthetic limbs through stump muscles.

Biomechatronics has seen 14) _____ growth with the development of advanced prosthetic limbs like the i-limb and the Power Foot BiOM. However, challenges remain, including high costs, limited insurance coverage, and mechanical limitations such as battery power and reliability. Progress in these areas is crucial to make biomechatronic technologies more 15)_____ and efficient for healthcare applications.

Academic Expression

Pair work: Discuss with your partner and compare the two possible paraphrases of each sentence and decide which one expresses the original meaning more academically.

1. Biomechatronics integrates biology and mechatronics, encompassing robotics and neuroscience.

   a. Biomechatronics integrates the fields of biology and mechatronics, encompassing the domains of robotics and neuroscience.

   b. Biomechatronics combines biology and mechatronics, including stuff like robots and neuroscience.

2. Biomechatronic devices have applications in prosthetic limbs and engineering solutions for respiration, vision, and the cardiovascular system.

   a. Biomechatronic devices are used for things like artificial limbs and engineering solutions for breathing, vision, and the heart.

   b. Biomechatronic devices find practical applications in the realm of prosthetic limbs and engineering solutions for respiratory, visual, and cardiovascular systems.

3. The functioning of biomechatronics mimics how the human body works, involving motor impulses, feedback from nerve cells, and muscle adjustments.

   a. Biomechatronics works like the human body, using signals from the brain, nerves, and muscles to do its thing.

   b. The operation of biomechatronics emulates the physiological functioning of the human body, incorporating elements such as motor impulses, feedback from neural cells, and muscular adaptations.

4. Biosensors detect user intentions and motions, relaying information to a controller.

   a. Biosensors perceive user intentions and movements, transmitting this data to a central controller.

   b. Biosensors pick up what the user wants to do and how they move, sending that info to a controller.

5. Mechanical sensors measure device information and interface with biosensors or controllers.

    a. Mechanical sensors measure device info and connect with biosensors or controllers.

    b. Mechanical sensors quantify device information and establish connections with biosensors or controllers.

6. The controller interprets user intentions and feedback information and controls the movements of the biomechatronic device.

    a. The controller deciphers user intentions and feedback, governing the movements of the biomechatronic device.

    b. The controller figures out what the user wants and how things are going, and makes the biomechatronic device move accordingly.

7. The actuator produces force and movement in response to control input.

    a. The actuator is what makes the device move and produce force based on the user's commands.

    b. The actuator generates force and motion in response to control input.

8. Neural interfacing allows biomechatronic devices to connect with the user's muscles and nerves, enabling bidirectional information exchange.

    a. Neural interfacing facilitates bi-directional information exchange between biomechatronic devices and the user's muscular and nervous systems.

    b. Neural interfacing allows the biomechatronic device to connect with the user's muscles and nerves, so they can exchange information back and forth.

Understanding the Text

Ⅰ. **Pair work**: Work with your partner and take turns asking and answering the following questions according to the information contained in the text.

1. What is the goal of biomechatronics?

2. How does biomechatronics mimic the human body?

3. What are biosensors used for in biomechatronics?

4. What is the role of mechanical sensors in biomechatronics?

5. What is the function of the controller in a biomechatronic device?

6. What is an actuator in biomechatronics?

7. Which areas are emphasized in current biomechatronics research?

8. What are some challenges facing biomechatronic technologies in the healthcare market?

II. Group work: Please work in groups and complete the structural table with the information from Text A.

Function	Section	Content
Introduction	A	Biomechatronics aims to 1) _____ biology and mechatronics and involves 2) _____ and neuroscience.
Body	B	Biosensors are used to detect user intentions and motions, 3) _____ information to a controller.
	C	Electromechanical sensors 4) _____ information about the biomechatronic device and relate it to the biosensor or 5) _____.
	D	The controller relays user intentions to actuators and 6) _____ feedback information from biosensors and mechanical sensors.
	E	The actuator 7) _____ force and movement, either as an artificial muscle or a motor that assists or 8) _____ the user's muscle.
	F	Current research focuses on 9) _____ human motions, studying the 10) _____ between electronic devices and the nervous system, and testing the use of living muscle tissue as actuators.
	G	Neural 11) _____ enables communication between biomechatronic devices and the user's muscle systems and nerves.

Continued Table

Function	Paragraph	Content
Conclusion	H	Biomechatronic devices have high demand but face challenges such as high costs, lack of 12)_____ in insurance policies, mechanical obstructions, and 13)_____ battery power and reliability.

Rise of a Great Power

Translate the following Chinese part into English and make it a complete English text.

Biomechanics is the discipline that studies the biomechanical characteristics and principles of motion in living organisms, and 1)(它对于理解生物体的结构、功能和运动机制具有重要意义)_____

_____.
Biomechanics also has significant applications in medical imaging, human simulation, and the development of medical devices.

2)(中国的科研人员一直在利用生物力学原理研究人体运动和姿势控制,探索运动障碍和康复治疗的有效方法)_____

_____. In addition, in-depth research has been conducted in fields such as cellular mechanics, biomaterials, and biomechanical simulation, 3)(推动了生物力学的发展)_____

_____.
They investigate cellular motion behavior, cellular mechanical properties, and the mechanical performance of biomaterials, 4)(为生物医学工程和再生医学的发展提供了重要的理论和实践基础)_____

_____.
In the future, 5)(中国将继续加强对生物力学领域的支持和投资,促进生物力学研究的创新和应用)_____

_____, further advancing scientific and technological progress, and enhancing human well-being.

Text B
Minimally Invasive Surgery and Surgical Robotics

Research Background

Minimally invasive surgery (MIS) and surgical robotics have evolved as alternatives to traditional open surgery, offering benefits such as reduced pain, shorter hospital stays, and improved outcomes. MIS gained popularity in the mid-1970s, utilizing small incisions and advanced imaging technology. The introduction of robot-assisted surgery in the 1980s further enhanced surgical capabilities. The da Vinci surgical robot became widely used but was faced limitations in dexterity and resolution. Teleoperation became the standard, allowing surgeons to remotely control surgical tools. Recent advances in telecommunication and control technologies have facilitated faster development in teleoperated robotic surgery. However, challenges remain, including the adoption of flexible robotic systems and addressing medical and ethical concerns.

Introduction Questions

· What are the advantages of minimally invasive surgery (MIS) compared to traditional open surgery?

· How has the integration of surgical robotics with MIS improved surgical outcomes?

· What are the challenges in adopting flexible robotic systems for MIS?

A. Studies on medical robotics and biomechatronic systems have been tracked back to the 1970s when open surgery, the traditional approach used for medical interventional, started to phase out. Typically, the orthodox approach, which dates back to the 1600s, involves accessing the internal organs via a large orifice to enhance safe manipulation of specific instruments and visualization of the procedures.

B. For instance, in cardiac interventions, the traditional approach involves opening the chest cavity widely, to perform surgery on valves, or vessels of the heart. Thus, open surgery is inherent with intense pain, surgical site infection, high hemorrhage, and long postsurgical hospital stays suffered by patients. However, use of anesthesia became a modern way to managing the traumatic pains associated with open-heart surgery in the mid-nineteenth century, while the antiseptic surgical methods, developed in 1860, was later encouraged to tackle the common surgical site infection problem.

C. In mid-1970s, minimally invasive surgery (MIS) became a better alternative to open surgery. Instead of the single large opening used in open surgery, intra-body access was gained through multiple invasions which are very small; therefore, reduced the blood loss and longer postsurgical hospital stays suffered by patients. However, surgeons experience loss of vision causing increased surgical time; thus, exposed to more orthopedic injuries. Coupled with advances in imaging technology from solid-state cameras and high-definition video displays in the 1980s, surgeons could view patients' anatomical and pathological views in high-quality 3-Dimensional images during surgery. Hence, loss of vision in MIS was eliminated.

D. Overtime, patients and surgeons steadily preferred the laparoscopic approach over the traditional open surgery. In addition to decreased postoperative morbidity and improved cosmesis, advantages of MIS became a motivation towards robot-assisted surgery in the 1980s when Arthrobot was used for surgical prostatectomies and cardiac valve repair.

E. A major motivation of robot-assisted MIS interventions is overcoming limitations of the conventional MIS approach, and enhancing the capabilities of surgeons when performing surgery. Thus, notion towards minimally invasive flexible surgery (MIFS) is devised as procedures involving proximal meandering of devices attached to laparoscope for surgical diagnosis and therapy in human. A technological roadmap showing some major developments towards flexible robotic systems in MIS is recalled from Vitiello et al.

F. Surgical robotics has evolved significantly in the past four decades, while rapt usage has triggered a paradigm shift with measurable positive impact in surgical outcomes. Moreover, integration of robotics with MIS has put forward better ways to eliminating some limitations of the traditional open surgery. While the da Vinci surgical robot (Intuitive Surgical Inc., California, CA, USA) remains the most widespread robotic system used for facilitating complex surgeries with minimal invasions, its performance is limited by proximal dexterity, resolution of human fine motion, reaction time, and cognitive skills. Sequel to its FDA approval in 2000, more than 5114 units have been globally installed and deployed for procedures, such as biopsy and surgery. In all cases, da Vinci robotic systems utilize minimal incision ports for insertion of surgical tools into patients while navigation of its end effector and actual surgical procedures are directed by physicians under teleoperated guidance.

G. Teleoperation is a standard for effective and safe MIS procedures in interventional surgery. In a typical MIS setup, surgeons sit at the master control station, located outside a surgical room, to issue control commands via communication channels. Control signals are sent to maneuver the surgical end effectors on the slave device, while visual/haptic data are feedback to the surgeon. This convention gives physicians greater capabilities in carrying out surgical procedures without necessarily having direct contacts with patients. A referable vantage of teleoperation in MIS is notable minimization in occupation-related injuries, such as orthopedic pains and longtime exposure to radiation. Teleoperated robotic systems have been commercially developed, and well reported with improved effectiveness during MIS procedures; however, adoption of flexible robotic systems into standard practice is still limited, globally. Studies have initiated the design of robotic systems that can precisely and desirously mimic surgeons' hand movements to facilitate complex surgical procedures through single-port minimal invasion.

H. Recent advances in telecommunication and intelligent control technologies have enhanced fast developments in teleoperated robotic surgery. Currently, fascinating surgical procedures can be carried out with master/slave (MS) control models. Existing schemes are based on direct, coordinated teleoperation, and supervisory control, while autonomous control with virtual or augmented presence is still being investigated. For autonomous surgery to be adopted for MIS, medical, legal, and ethical concerns are yet to be standardized. Also, visual and haptic appliances are required to complement the direct vision and tactile feedbacks which are inherent with the traditional open surgery.

I. Despite the prospects of robots that have been developed for MIFS, applications of flexible prototypes with redundant snake-like or continuum structures are still globally limited. This can be attributed to a wide range of issues, such as lack of efficient constraints modeling for motion control and teleoperation of the robotic systems.

Vocabulary List

biomechatronic
  relating to the integration of biology and mechatronics 生物力学的

anesthesia

the use of medication to induce a loss of sensation or consciousness during surgery or medical procedures 麻醉

antiseptic

relating to substances that prevent the growth of disease-causing microorganisms 防腐的

orthopedic

relating to the branch of medicine dealing with the correction of deformities of bones or muscles 骨科的

laparoscopic

relating to a surgical procedure performed through small incisions with the aid of a camera 腹腔镜的

cardiac valve

a valve located within the heart 心脏瓣膜

haptic

relating to the sense of touch or the perception of tactile sensations 触觉的

tactile

of or connected with the sense of touch 触觉的

Language Enhancement

Ⅰ. Complete the following sentences with words listed in the box below. Change the form where necessary.

deploy	invasive	cognitive	resolution	alternative
redundant	install	diagnose	fascinate	enhance
integration	trigger	flexible	appliance	access
facilitate	vantage	meandering	interventional	complement

1. The _____ components in the system serve as a backup to ensure uninterrupted operation.

2. The use of virtual reality technology _____ the traditional learning methods in medical education.

3. The robotic _____ assisted the surgeon in performing precise and delicate procedures.

4. The intricacies of the human brain continue to _____ neuroscientists worldwide.

5. Having a _____ point from above allows the researchers to observe the behavior of marine creatures more effectively.

6. The military decided to _____ the latest surveillance drones to monitor the enemy's activities.

7. The technicians are working diligently to _____ the new software on all the computers in the laboratory.

8. _____ computing systems are revolutionizing the way we process and analyze complex data.

9. The high-_____ imaging technology provides detailed visualization of internal structures.

10. The new software update aims to _____ the performance and functionality of the device.

11. The loud noise _____ an immediate response from the automated emergency system.

12. The _____ of artificial intelligence and healthcare systems promises more efficient patient care.

13. The _____ robotic arm can adapt to various surgical procedures with ease.

14. The doctor used advanced imaging techniques to accurately _____ the patient's condition.

15. The security clearance granted him _____ to the highly classified research facility.

16. The new software tool aims to _____ communication and collaboration among team members.

17. Seeking _____ treatment options has become a common practice in modern medicine.

18. The river's _____ path created challenges for the construction of an interventional facility.

19. The _____ surgical procedure was necessary to remove the tumor and save the patient's life.

20. Doctors are increasingly utilizing _____ techniques to perform minimally invasive procedures using advanced medical technology.

Ⅱ. Complete the following sentences with phrases listed in the box below. Change the form where necessary.

associate with	put forward to	carry out	attribute to	be inherent with
phase out	contact with	expose to	encourage to	track back to

1. The cybersecurity team was able to _____ the source of the cyber attack using advanced tracing techniques.

2. The company plans to _____ the outdated technology and replace it with more advanced and efficient systems.

3. Artificial intelligence _____ the ability to learn and adapt based on data and algorithms.

4. The new software update _____ improved performance and enhanced user experience.

5. Employees _____ participate in training programs to stay updated with the latest technological advancements.

6. The research study _____ participants _____ various stimuli to measure their cognitive responses.

7. The scientist _____ a hypothesis _____ explain the observed phenomenon in the field of quantum computing.

8. The team had frequent _____ experts in the industry to gain insights and guidance for their project.

9. The research team _____ extensive experiments to validate their findings and conclusions.

10. The success of the project can _____ the collaborative efforts of the entire team.

Ⅲ. Complete the summary using the suitable words in Text B.

Minimally invasive surgery (MIS) and surgical robotics have 1)_____ the field of medical interventions, offering alternative approaches to traditional open surgery. MIS, introduced in the mid-1970s, involves 2)_____ internal organs through small incisions, reducing pain, blood loss, and hospital stays for patients. 3)_____ in imaging technology have improved visualization during surgery. Over time, laparoscopic approaches gained 4)_____ over open surgery, leading to the emergence of robot-assisted surgery in the 1980s. Robotic systems, like the da Vinci surgical robot, have further 5)_____ MIS outcomes, although limitations in dexterity and 6)_____ skills still exist. Teleoperation plays a crucial role in MIS, allowing surgeons to remotely control robotic devices for 7)_____ procedures. However, the adoption of flexible robotic systems is limited.

Recent developments in telecommunication and control technologies have 8) _____ the progress of teleoperated robotic surgery, with the potential for autonomous control being explored. However, standardization of medical, legal, and ethical concerns is necessary for widespread acceptance. Despite the 9) _____ of flexible prototypes, issues such as motion control and teleoperation constraints modeling need to be 10) _____ for their global application.

Academic Expression

Pair work: Discuss with your partner and compare the two possible paraphrases of each sentence and decide which one expresses the original meaning more academically.

1. Studies on medical robotics and biomechatronic systems have been tracked back to the 1970s when open surgery, the traditional approach used for medical interventional, started to phase out.

    a. Research on medical robotics and biomechatronic systems goes way back to the 1970s when they started moving away from open surgery, which used to be the go-to method for medical intervention.

    b. Studies on medical robotics and biomechatronic systems can be traced back to the 1970s, when open surgery, the conventional approach employed in medical intervention, began to be phased out.

2. In mid-1970s, minimally invasive surgery (MIS) became a better alternative to open surgery.

    a. During the mid-1970s, minimally invasive surgery (MIS) emerged as a more favorable alternative to open surgery.

    b. Around the 1970s, doctors found that minimally invasive surgery (MIS) was a better option than open surgery.

3. Coupled with advances in imaging technology from solid-state cameras and high-definition video displays in the 1980s, surgeons could view patients' anatomical and pathological views in high-quality 3-Dimensional images during surgery.

    a. In the 1980s, fancy imaging technology like solid-state cameras and high-definition video displays came along, allowing surgeons to see clear 3D images of what's going on inside patients' bodies during surgery.

    b. With the advent of advanced imaging technology, including solid-state cameras and high-definition video displays in the 1980s, surgeons gained the ability to visualize patients' anatomical and pathological conditions in high-quality three-dimensional images during surgical procedures.

4. Overtime, patients and surgeons steadily preferred the laparoscopic approach over the traditional open surgery.

  a. Over time, laparoscopic procedures gradually gained preference over traditional open surgery among both patients and surgeons.

  b. As time went on, more and more patients and surgeons preferred laparoscopic procedures instead of the old-school open surgery.

5. A major motivation of robot-assisted MIS interventions is overcoming limitations of the conventional MIS approach and enhancing the capabilities of surgeons when performing surgery.

  a. Surgeons got excited about using robots to make MIS even better and overcome the limitations of regular MIS procedures.

  b. The integration of robotic technology into MIS procedures aims to address the limitations of conventional MIS approaches and augment surgeons' capabilities during surgery.

6. Surgical robotics has evolved significantly in the past four decades, while rapt usage has triggered a paradigm shift with measurable positive impact in surgical outcomes.

  a. The field of surgical robotics has witnessed significant advancements over the past four decades, leading to a paradigm shift that has positively impacted surgical outcomes.

  b. Surgical robotics has come a long way in the last few decades and has made a big difference in how surgeries turn out.

7. In all cases, da Vinci robotic systems utilize minimal incision ports for insertion of surgical tools into patients while navigation of its end effector and actual surgical procedures are directed by physicians under teleoperated guidance.

  a. The da Vinci robotic systems are super popular and use small incisions to put surgical tools inside patients, and the doctors control everything remotely.

  b. Da Vinci robotic systems, which have been extensively utilized, rely on minimal incisions for the insertion of surgical instruments and are operated by physicians under teleoperated guidance.

8. Teleoperated robotic systems have been commercially developed and well reported with improved effectiveness during MIS procedures; however, adoption of flexible robotic systems into

standard practice is still limited, globally.

a. Teleoperated robotic systems have been developed and proven effective in MIS procedures, although the global adoption of flexible robotic systems in standard practice remains limited.

b. Teleoperated robotic systems have been developed and tested in MIS procedures and have shown to be pretty effective, but not everyone is using flexible robots all around the world.

Understanding the Text

Ⅰ. **Text B has nine paragraphs, A-I. Choose the correct heading for paragraphs A-I from the list of headings below.**

List of headings

Ⅰ	The Impact of Robotics in Other Industries
Ⅱ	Advancements in Surgical Robotics and Integration
Ⅲ	Advances in Telecommunication and Control Technologies
Ⅳ	Innovations in Minimally Invasive Flexible Surgery
Ⅴ	Teleoperation in Interventional Surgery
Ⅵ	Shift towards Laparoscopic and Robot-Assisted Surgery
Ⅶ	The History and Development of Medical Imaging Technologies
Ⅷ	Challenges of Open Surgery and Advancements
Ⅸ	Medical Robotics and Biomechatronic Systems
Ⅹ	Minimally Invasive Surgery and Imaging Technology
Ⅺ	Limitations and Future Directions

Paragraph A:
Paragraph B:
Paragraph C:
Paragraph D:
Paragraph E:
Paragraph F:
Paragraph G:
Paragraph H:
Paragraph I:

Ⅱ. Do the following statements agree with the information given in Text B? Write your judgment.

TRUE	if the statement agrees with the information
FALSE	if the statement contradicts the information
NOT GIVEN	if there is no information on this

1. Studies on medical robotics and biomechatronic systems can be traced back to the 1970s.

2. The traditional approach used in medical interventions involves accessing internal organs through a small opening.

3. Open surgery is associated with intense pain, surgical site infection, high hemorrhage, and long postsurgical hospital stays.

4. Anesthesia was introduced in the mid-nineteenth century to manage the pain associated with heart surgery.

5. Minimally invasive surgery (MIS) was introduced in the mid-1970s as an alternative to open surgery.

6. Surgeons using MIS will never experience any loss of vision during surgical procedures.

7. The integration of robotics with MIS has eliminated the loss of vision in surgical procedures.

8. Teleoperation is a standard practice in interventional surgery and has minimized occupation-related injuries.

Ⅲ. Choose the best answer according to Text B.

1. What was the traditional approach used for medical intervention before minimally invasive surgery (MIS) emerged?

A) Robotic surgery

B) Open surgery

C) Laparoscopic surgery

D) Teleoperated surgery

2. What motivated the development of robot-assisted surgery in the 1980s?

A) Decreased postoperative morbidity

B) Improved cosmesis

C) Limitations of conventional MIS approach

D) Standardization of medical, legal, and ethical concerns

3. What is a limitation of the da Vinci surgical robot?

A) Lack of proximal dexterity

B) Lack of teleoperated guidance

C) Lack of minimal incision ports

D) Lack of visual and haptic feedback

4. What is a benefit of teleoperation in minimally invasive surgery (MIS)?

A) Enhanced visualization of procedures

B) Reduced surgical site infection

C) Greater capabilities for surgeons without direct patient contact

D) Improved resolution of human fine motion

5. What is one of the issues limiting the global application of flexible robotic systems for minimally invasive surgery (MIS)?

A) Lack of fast developments in telecommunication

B) Lack of efficient constraints modeling for motion control

C) Lack of supervisory control in MS control models

D) Lack of direct vision and tactile feedback in autonomous surgery

Rise of a Great Power

Translate the following Chinese text into English.

1. (微创手术和外科机器人技术是当代外科医学领域的重要发展方向) _____

_____. Minimally invasive surgery reduces intraoperative bleeding and tissue damage by using smaller incisions and specialized surgical instruments. 2)(中国的医学研究机构和医院积极探索并采用了微创手术技术) _____

_____, such as laparoscopic surgery and thoracoscopy. 3)(中国的医生和科研人员也致力于改进微创手术技术,以提高手术的精确性和安全性) _____

_____. China has developed advanced surgical robotic systems 4)并在临床实践中进行了验证) _____

_____. Surgical robotic systems provide surgeons with precise surgical manipulation, enhanced visualization, and operational control, while minimizing surgical trauma. Additionally, 5)(中国的科研人员还致力于改进外科机器人系统的性能和功能)_____

_____
_____ to meet the demands of different surgical procedures.

3-1　Unit Three-Answer and Translation

Unit Four Electromechanics

Warm-up

Currently, the world economic and industrial pattern is undergoing major adjustments, changes, and developments. On the one hand, the impact of the international financial crisis continues, economic recovery is slow, and there are increasing uncertainties in development; on the other hand, a new round of technological revolution and industrial transformation is brewing globally, especially the deep integration of new-generation information technology and manufacturing, coupled with breakthroughs in new energy, new materials, biotechnology, and other fields, which are causing far-reaching industrial changes.

Thought-provoking Questions

· In the context of the ongoing major adjustments and developments in the world economic and industrial pattern, how can mechanical engineering professionals adapt to the changing landscape and leverage the opportunities presented by the new round of technological revolution and industrial transformation?

· With the deep integration of new-generation information technology and manufacturing, what potential challenges and opportunities do you foresee in the field of mechanical engineering?

· How can you, as a mechanical engineering student, prepare yourself to address these challenges and seize the opportunities?

Text A
Evolution and Applications of Electromechanics

Research Background

Electromechanics combines electrical engineering and mechanical engineering, focusing on the interaction between electrical and mechanical systems. It encompasses devices that convert electrical signals into mechanical movement or vice versa. Electromechanical devices were widely used in early systems like typewriters, teleprinters, clocks, and early computers. The development of electromechanics accelerated during the Industrial Revolution and both World Wars. Post-war, electromechanical systems found applications in military equipment and household appliances. However, the rise of solid-state electronics led to a decline in electromechanics. Today, power companies mainly use electromechanical processes, while most consumer devices rely on integrated microcontroller circuits. Piezoelectric devices, which convert electrical signals into sound or mechanical vibration, are also considered electromechanical.

Introduction Questions

· What is electromechanics and how does it combine electrical and mechanical engineering?

· What are some examples of electromechanical devices?

· How has the development of solid-state electronics impacted electromechanics?

A. Electromechanics is an interdisciplinary field that combines concepts and techniques from electrical engineering and mechanical engineering. It focuses on the interaction between electrical and mechanical systems, exploring how these two domains interact and function together. This integration of electrical and mechanical processes has led to the development of various electromechanical devices and systems. In this discussion, we will delve into the principles, historical significance, and applications of electromechanics.

B. Electromechanics involves the integration of electrical and mechanical processes, drawing from the fields of electrical engineering and mechanical engineering. It examines the holistic interaction between electrical and mechanical systems, particularly evident in the operation of DC or AC rotating electrical machines. These machines can either generate power through mechanical processes (generator) or utilize electrical power to produce mechanical effects (motor). This broader context of electrical engineering also encompasses electronics engineering.

C. Electromechanical devices encompass components that incorporate both electrical and mechanical functionalities. While a manually operated switch can be considered electromechanical due to its mechanical movement leading to an electrical output, the term typically refers to devices that utilize electrical signals to create mechanical movement or convert mechanical movement into electrical signals. Electromechanical devices often rely on electromagnetic principles, such as relays that enable voltage or current control of isolated circuits through mechanical switching of contacts, and solenoids that actuate moving linkages, as seen in solenoid valves.

D. In the past, electromechanical devices played a significant role in complex subsystems across various domains. Electric typewriters, teleprinters, clocks, early television systems, and even the first digital computers heavily relied on electromechanical components. However, the advent of solid-state electronics has led to the replacement of many electromechanical devices in numerous applications.

E. The first electric motor was invented in 1822 by Michael Faraday, shortly after Hans Christian Orsted's discovery of the relationship between electric current and magnetic fields. Faraday's motor consisted of a wire partially submerged in mercury, with a magnet at the bottom. When the wire was connected to a battery, the interaction between the magnetic field generated by the magnet and the current in the wire caused it to spin.

F. Ten years later, in another breakthrough, Michael Faraday invented the first electric generator. This device involved a magnet passing through a coil of wire, inducing a current that could be measured using a galvanometer. Faraday's research and experiments laid the foundation for many modern electromechanical principles.

G. The growing interest in electromechanics gained momentum alongside the advancements in long-distance communication during the Industrial Revolution. The increased demand for intracontinental communication drove the integration of electromechanical systems into public services. For instance, relays were initially used in telegraphy to regenerate telegraph signals. Early telephone exchanges employed devices like the Strowger switch, the Panel switch, and crossbar switches. The adoption of crossbar switches became widespread in the mid-20th century, initially in countries like Sweden, the United States, Canada, and Great Britain, later spreading globally.

H. The progress of electromechanical systems accelerated from 1910 to 1945 due to the two world wars. During World War I, spotlights and radios utilizing electromechanics were widely employed by all countries. By World War II, nations had centralized and developed their military capabilities around the versatility and power of electromechanical systems. An example of this is the alternator, initially designed to power military equipment in the 1950s, which later found applications in automobiles in the 1960s. Post-war America experienced significant advancements in electromechanical systems, with household chores being revolutionized by the introduction of electromechanical devices such as microwaves, refrigerators, and washing machines. However, electromechanical television systems from the late 19th century were less successful compared to other developments.

I. Electric typewriters, also known as "power-assisted typewriters," emerged and evolved until the 1980s. They featured a single electrical component—the motor. Instead of directly moving the typebar with each keystroke, the motor engaged mechanical linkages to transfer mechanical power to the typebar. This principle was also applied in the later IBM Selectric typewriters. In 1946, Bell Labs developed the Bell Model V computer, an electromechanical relay-based device with slow processing cycles. In 1968, electromechanical systems were still under consideration for aircraft flight control computers until the adoption of large-scale integration electronics in the Central Air Data Computer.

J. In contemporary times, electromechanical processes primarily find applications in power companies. Mechanical movement is converted to electrical power by fuel-based generators. Additionally, renewable energy sources such as wind and hydroelectric systems utilize mechanical processes that convert movement into electricity.

K. Over the last three decades of the 20th century, electromechanical devices gradually became less prevalent as cost-effective alternatives emerged. The decline in their usage can be attributed to the rise of integrated microcontroller circuits, which incorporate millions of transistors and execute tasks through logical programming. Unlike electromechanical components with moving parts, these electronic circuits are highly reliable and immune to mechanical wear and eventual failure. As a result, electronic circuits without moving parts have become integral to various systems, including traffic lights and washing machines, and are commonly used in simple feedback control systems.

L. In conclusion, electromechanics combines principles from electrical engineering and mechanical engineering to examine the interaction between electrical and mechanical systems, which has played a crucial role in various applications throughout history, from early telegraphy and telephone exchanges to the development of electric motors and generators. However, the advent of solid-state electronics has led to the replacement of many electromechanical devices. Nonetheless, electromechanical processes still find utility in power generation, particularly with fuel-based generators and renewable energy systems. With the emergence of integrated microcontroller circuits, the reliance on electromechanical devices has diminished, as electronic circuits without moving parts offer enhanced reliability. Nevertheless, the fundamental principles and contributions of electromechanics continue to shape our modern world.

Vocabulary List

electromechanics
an interdisciplinary field that combines concepts and techniques from electrical engineering and mechanical engineering　电机学

relays
devices that allow control of electrical circuits by using an electromagnet to operate a switch mechanism　继电器;转发器;中继器

solenoid
a coil of wire that acts as an electromagnet when an electric current passes through it　电磁铁;线圈;细长螺线管

actuate
to cause something to start functioning or put into action　驱动;启动

telegraphy

the use or operation of the telegraph system for sending and receiving messages at a distance by means of electrical signals 电报术；电信术

crossbar

a device with several bars or wires crossing each other that provides connections between different points in an electrical circuit 横杆；十字形杆；十字开关

typebar

a metal bar with a raised character on the end, used in typewriters to print the corresponding letter or symbol on paper 打字机键杆

transistor

a semiconductor device used to amplify or switch electronic signals and electrical power 晶体管；晶体管放大器；三极管

Language Enhancement

Ⅰ. Complete the following sentences with words listed in the box below. Change the form where necessary.

encompass	linkage	evolve	prevalent	diminish
utilize	actuate	engage	enhance	versatility
integration	incorporate	incorporate	emerge	advent

1. _____ of various technologies is essential for the development of advanced smart cities.

2. Scientists _____ nanotechnology to create more efficient solar cells.

3. The design of the new smartphone _____ cutting-edge features and functionalities.

4. The robot _____ artificial intelligence and machine learning algorithms for autonomous decision-making.

5. The motion sensor _____ the lights when it detects movement in the room.

6. The _____ between the software and hardware components is crucial for the smooth operation of the computer system.

7. The _____ of 3D printing has revolutionized the manufacturing industry.

Unit Four Electromechanics

8. The _____ of smartphones allows users to perform a wide range of tasks, from communication to multimedia consumption.

9. With the advancement of technology, new forms of communication tools _____ regularly.

10. Technology continues to _____ rapidly, bringing innovative solutions to everyday problems.

11. It is important to _____ students in hands-on experiments to foster their interest in science and technology.

12. Artificial intelligence is becoming more _____ in our daily lives, from virtual assistants to autonomous vehicles.

13. The design of the new car _____ eco-friendly materials and energy-efficient systems.

14. The use of renewable energy sources helps _____ our dependence on fossil fuels.

15. Advanced algorithms _____ the accuracy and speed of data analysis in scientific research.

Ⅱ. **Complete the following sentences with phrases listed in the box below. Change the form where necessary.**

convert into	transfer to	be immune to	delve into
draw from	drive into	consist of	refer to

1. Engineers _____ multiple disciplines to design innovative solutions.
2. The term "artificial intelligence" _____ the development of intelligent machines.
3. The software _____ various modules that work together to provide a seamless user experience.
4. The new electric car _____ the future of sustainable transportation.
5. Data _____ the cloud for secure storage and easy accessibility.
6. Heat energy _____ electricity through the process of thermoelectric conversion.
7. Researchers _____ the complexities of quantum mechanics to unlock its potential applications.
8. The new material _____ corrosion, making it ideal for harsh environments.

Ⅲ. **Choose the most suitable words to complete the summary of Text A. Change the form where necessary.**

encompass	extensive	range	prominence	generator
convert	transform	examine	advent	appliance

Continued Table

decline	utilizing	shape	specialize	utilize
integral	invention	foster	merge	significant

Electromechanics, an interdisciplinary field 1) _____ electrical engineering and mechanical engineering, 2) _____ the interaction between electrical and mechanical systems. It finds applications in rotating electrical machines, such as 3) _____ and motors, which 4) _____ mechanical energy to electrical power or vice versa. Electromechanical devices 5) _____ both electrical and mechanical processes, 6) _____ from relays to solenoids. While electromechanics had a 7) _____ role in early subsystems like typewriters and teleprinters, solid-state electronics replaced many applications. Notable milestones include Michael Faraday's 8) _____ of the electric motor and generator in the 19th century. Electromechanics gained 9) _____ during the World Wars and found 10) _____ use in spotlights, radios, and military equipment. Post-war, electromechanical systems 11) _____ households with 12) _____ like microwaves and washing machines. However, the 13) _____ of integrated microcontroller circuits led to the 14) _____ of electromechanical devices, as electronic circuits offered greater reliability. Nevertheless, electromechanical processes remain 15) _____ to power generation, with fuel-based generators and renewable energy systems 16) _____ mechanical movement. In conclusion, electromechanics continues to 17) _____ its principles to 18) _____ power generation and 19) _____ domains, while also 20) _____ the synergy between electrical and mechanical engineering.

Academic Expression

Pair work: Discuss with your partner and compare the two possible paraphrases of each sentence and decide which one expresses the original meaning more academically.

1. Electromechanical devices are ones which have both electrical and mechanical processes.

　　a. Electromechanical devices are the ones that combine electrical and mechanical processes.

　　b. Electromechanical devices encompass devices that involve both electrical and mechanical processes.

2. Often involving electromagnetic principles such as relays, which allow a voltage or current to control another circuit voltage or current by mechanically switching sets of contacts.

　　a. Frequently incorporating electromagnetic principles, relays enable the control of one circuit's voltage or current by another through mechanical contact switching.

  b. Many times, electromagnetic principles are involved, like in relays, which let voltage or current control another circuit's voltage or current by switching contacts mechanically.

3. The first electric motor was invented in 1822 by Michael Faraday.

  a. It was Michael Faraday who came up with the first electric motor in 1822.

  b. Michael Faraday is credited with the invention of the first electric motor in 1822.

4. Electromechanical systems saw a massive leap in progress from 1910 to 1945 as the world was put into global war twice.

  a. The period between 1910 and 1945 witnessed a significant advancement in electromechanical systems due to two global wars.

  b. From 1910 to 1945, electromechanical systems made huge progress because of the two global wars.

5. Properly designed electronic circuits without moving parts will continue to operate correctly almost indefinitely and are used in most simple feedback control systems.

  a. Electronic circuits that are designed properly and don't have any moving parts can work perfectly for a very long time, and they are used in most basic feedback control systems.

  b. Well-designed electronic circuits that lack moving parts are capable of long-term correct operation and find application in the majority of basic feedback control systems.

6. The Industrial Revolution's rapid increase in production gave rise to a demand for intracontinental communication, allowing electromechanics to make its way into public service.

  a. The rapid surge in production during the Industrial Revolution created a need for intracontinental communication, paving the way for the utilization of electromechanics in public services.

  b. During the Industrial Revolution, with the massive increase in production, there was a growing demand for communication within continents, which led to the adoption of electromechanical systems in public services.

7. Relays originated with telegraphy as electromechanical devices were used to regenerate telegraph signals.

a. Relays have their roots in telegraphy, where electromechanical devices were used to regenerate telegraph signals.

b. The origins of relays can be traced back to telegraphy, where electromechanical devices were employed for the purpose of signal regeneration.

8. Post-war America greatly benefited from the military's development of electromechanics as household work was quickly replaced by electromechanical systems such as microwaves, refrigerators, and washing machines.

a. The development of electromechanics by the military during the post-war period had a significant impact on American society, leading to the rapid replacement of household chores with electromechanical systems like microwaves, refrigerators, and washing machines.

b. After the war, the advancement of electromechanics by the military brought great benefits to America, as it led to the quick adoption of household appliances such as microwaves, refrigerators, and washing machines, which made chores much easier.

Understanding the Text

Ⅰ. **Pair work**: Work with your partner and take turns asking and answering the following questions according to the information contained in the text.

1. What is electromechanics?

2. What are some examples of electromechanical devices?

3. What was the significance of Michael Faraday's inventions?

4. How did the shift to solid-state electronics impact electromechanical devices?

5. What are some applications of electromechanics?

6. What are some historical developments in electromechanics?

7. How do integrated microcontroller circuits differ from electromechanical devices?

8. What is the relationship between electromechanics and solid-state electronics?

___

II. Group work: Work in groups and complete the following analysis of the structure with the information from Text A.

Function	Paragraph	Main Content
Introduction	A	Introduction to electromechanics as an interdisciplinary field that 1)_____ concepts and techniques from electrical engineering and mechanical engineering, 2)_____ to the development of various electromechanical devices and systems.
Body	B	Explanation of how electromechanics integrates electrical and mechanical processes, particularly in rotating electrical machines that 3)_____ power or produce mechanical effects.
	C	Discussion of electromechanical devices that incorporate both electrical and mechanical functionalities, 4)_____ electrical signals for mechanical movement or converting mechanical movement into electrical signals.
	D	Explanation of the historical significance of electromechanical devices in various 5)_____, their 6)_____ on solid-state electronics, and the replacement of many electromechanical devices in different applications.
	E	Description of Michael Faraday's invention of the first electric motor in 1822 and its operation based on the 7)_____ between magnetic fields and electrical current.
	F	Overview of Michael Faraday's invention of the first electric generator, which 8)_____ inducing a current in a wire coil using a magnet, and how it laid the 9)_____ for electromechanical principles.
	G	Discussion of the integration of electromechanical systems into public services, particularly in telegraphy and telephone 10)_____, and the use of relays and switches in communication systems during the Industrial Revolution.

Continued Table

Function	Paragraph	Main Content
Body	H	Explanation of the 11)_____ progress of electromechanical systems during the two world wars, their military applications, and their 12)_____ on post-war household devices.
Body	I	Description of electric typewriters and relay-based computers as examples of electromechanical devices, their eventual replacement by electronic circuits, and the 13)_____ of large-scale integration electronics in flight control computers.
Body	J	Explanation of the 14)_____ application of electromechanical processes in power generation, including fuel-based generators and 15)_____ energy systems that utilize mechanical processes to convert movement into electricity.
Body	K	Discussion of the 16)_____ in the usage of electromechanical devices due to the emergence of cost-effective alternatives like 17)_____ microcontroller circuits, which offer 18)_____ reliability compared to electromechanical components with moving parts.
Conclusion	L	Highlighting the interdisciplinary nature of electromechanics, its role in various applications throughout history, the 19)_____ of electromechanical devices by solid-state electronics, and the continued 20)_____ of electromechanical processes in power generation.

Rise of a Great Power

Translate the following Chinese part into English and make it a complete English text.

China's electrical engineering is a rapidly developing field 1)(涵盖了电气工程和机械工程的交叉领域)_____

_____. Against the backdrop of rapid development in information technology and manufacturing, 2)(中国的电机工程学在多个领域展现出巨大的潜力和前景)_____

_____.

The research and application of electrical engineering in China cover a wide range of areas, including electric motors, generators, sensors, control systems, and automation technologies. 3)(特别是在电动汽车和可再生能源领域,中国取得了显著的成就)_____
_____
_____. The flourishing electric vehicle market in China 4)(推动了电机工程学的研究和创新)_____
_____
_____. Additionally, 5)(在太阳能和风能等可再生能源领域也取得了重要进展)_____
_____
_____, where electrical engineering plays a pivotal role.

Text B
Core Principles for Flexible Surgical Robotics in MIS

Background Summary

Flexible robots have revolutionized minimally invasive surgery (MIS) by allowing surgeons to access and manipulate hard-to-reach organs. However, there are limited prototypes that meet the requirements for natural or single incised port procedures. Spatially flexible robots hold promise for delivering effective treatment in radiosurgery for cancerous organs. Core principles, including flexible navigation, single-port access, increased range of motion, spatial steerability, and surgical precision, are crucial for distinguishing flexible robots during MIS.

Introduction Questions

· What are the core principles for flexible robots in minimally invasive surgery?

> • How does single-port access contribute to the advancement of intra-body robotic surgery?
> • Why is surgical precision important in robotic surgery?

A. Flexible robots enable surgeons to visualize and manipulate anatomical organs that are difficult to access with previous the robots used in MIS. While several studies have been dedicated on the design and development of control systems for flexible robotic surgery, only a few prototypes are found useful for the procedures performed via natural or single incised port. For radiosurgery, it is presumed that oncologists can soon utilize spatially flexible robots in delivering large dosage of morphological and functional data for the treatment of cancerous organs in the human body. Omisore et al. proposed a flexible robotic system for minimally invasive radiosurgery of gastrointestinal tumors. The system uses redundant mechanisms for flexible navigation. Hence, we suggest that the subsequent core principles can be used to distinguish flexible robots during MIS procedures.

B. Single-Port Access Surgery: Single-port access is a revolutionary point towards intra-body robotic surgery. This core value enables surgeons to perform complex procedures on specific organs/tissues of the body through spatially flexible and narrow access. Thus, it is proposed as an important feature for flexible access surgery. It is characterized by navigating modular instruments or slender tools with very minimal invasion into the cavity of tubular organs or cannula with good cosmesis. As a result, patients can return to their activities of daily living soon as the procedure is completed. However, to avoid collisions in surgical workspace, distally actuated devices are utilized for proper triangulation along anatomical pathways or around targeted organs in the human body. Hence, surgeons can navigate the instrument along flexible pathways through a single port.

C. Range of Motion: Joint limitation is a common feature that bounds the accessible area of conventional arm robots during MIS. On the contrary, flexible surgical robots are characterized with increased range of motion base on a chain of serial links built with short interspaced actuations. Taking motion dexterity as a core feature that enhances spatial motion and maneuverability in flexible surgical robotics, unique parts of the robot must have an interconnecting joint designed with little to no offset and acute-to-obtuse extensibility. The effect of limited range of motion at each joint is enhanced per capita link added to the robotic structure. An ill-feature of this design principle is modular underactuation but the little contribution from each joint can substantially enhance the robot's pose, as desired. In flexible endoscopic robots,

HD cameras, with magnifying views, are utilized for see more capability during surgery; however, the research area is still open in flexible intravascular robots.

D. Spatial Steerability: Blurring out line of sights is quite difficult with conventional surgical robots because surgeon operates handheld instruments with limited tip dexterity. Aside this, single-port procedures require robotic devices whose unique link modules are capable of spatial movements. Each module can be well-fitted along curvy pathways and controlled along unique motion axis. Thus, modular structuring is another core principle we establish for flexible robots to be applicable for single-port intraluminal surgery. This third core feature can be characterized as a reach more capability of the emerging flexible surgical robotic systems that enhances surgeons to achieve highly spatial dexterity needed to reach more complex areas in patients. In this case, specialized actuation strategy is important to enable spatial flexibility and stable navigation of the manipulator.

E. Surgical Precision: For safety and reliability, precision is of great importance in robotic surgery. Accordingly, another important core principle proposed for flexible robotics used in MIS is surgical precision. This index can be used to quantify and analyze firmness of the flexible poses of the mechanism of appended surgical tools either when it is at stationary state or when moving along anatomical pathways. For objective analysis, this can be estimated with respect to the robotP's navigation and degree of dynamic stability during MIS. Flexible robotics has been previously described, in the literature, as a better means to performing surgical tasks with geometric accuracy and precision, as each module can be sensorized for guided motion. This is because, flexible robotic systems cannot be fully autonomous but rather, it will always be under surgeon's direction, allowing them to have full control during interventional procedures.

F. In essence, the above mentioned four core principles can be regarded as standards to adjudge the ability of a robotic device to navigation along a flexible anatomical path with desired trajectory for better surgical outcomes. Nonetheless, safe and efficient manipulation of the robots satisfying the four principles requires intelligent constraint control models to be applied for precise and timely intrabody navigation.

Vocabulary List

oncologist
 a doctor who specializes in the treatment of cancer 肿瘤学家;肿瘤科医生

morphological

relating to the form or structure of an organism　形态学的；形态的

gastrointestinal

relating to the stomach and intestines　胃肠的；胃肠道的

triangulation

the process of determining the location of a point by measuring angles to it from known points　三角定位；三边测量

underactuation

the limited control or actuation of a mechanism　动作不足；不充分激活

steerability

the ability to be steered or controlled, especially in terms of a vehicle or device　可操控性；可控性；可操纵性；可导向性

single-port intraluminal

a surgical procedure performed through a single situated or occurring within the lumen or cavity of a tubular organ or structure　腔内的；管腔内的

radiosurgery

a surgical technique that uses radiation for treatment　放射治疗；放射外科手术

trajectory

the curved path of an object moving through space　轨迹；弹道

maneuverability

the quality or ability to maneuver or control movement　操纵性；机动性

intravascular

situated or occurring within a blood vessel　血管内的；血管内注射的

dexterity

skill in performing tasks, especially with the hands　灵巧；敏捷

Language Enhancement

I. Complete the following sentences with words listed in the box below. Change the form where necessary.

utilize	acute	slender	dedicate	blurring
spatial	dynamic	adjudge	obtuse	presume
invasive	collision	dexterity	distinguish	redundant
interspace	magnify	manipulate	curvy	interconnecting

1. The _____ nature of the software allows for real-time adjustments and updates.
2. The judges will _____ the winner based on the performance and innovation of the technology.
3. The surgeon's _____ is crucial in performing intricate robotic surgeries.
4. The _____ design of the car enhances its aerodynamic efficiency.
5. _____ awareness is important in designing efficient layout for a factory.
6. The microscope can _____ the microscopic details of the specimen.
7. The _____ effect in the image can be reduced by using advanced image processing techniques.
8. The _____ angle of the robot's arm allows it to access tight spaces during surgery.
9. The _____ angle of the camera provides a wider field of view.
10. The _____ cables ensure seamless communication between different components of the system.
11. The _____ between the gears needs to be adjusted to avoid collisions.
12. The _____ design of the drone allows it to maneuver through narrow spaces.
13. It is important to _____ between different types of data to ensure accurate analysis.
14. The _____ backup systems ensure uninterrupted operation in case of failures.
15. _____ procedures should be minimized and replaced with non-invasive alternatives whenever possible.
16. Scientists _____ advanced algorithms to analyze large datasets.
17. Engineers can _____ the variables to optimize the performance of the system.
18. The team is _____ to developing innovative solutions for renewable energy.
19. Based on the initial data, we can _____ that the experiment will yield positive results.

20. The high-speed camera captures the fast-moving object with such speed that it results in a _____ effect in the recorded video.

II. Complete the summary with words in the Text B. Change the form where necessary.

This article 1)_____ the application of flexible robots in minimally 2)_____ surgery. It discusses the visualization and 3)_____ capabilities of flexible robots during surgical procedures, as well as their 4)_____ in radiosurgery and intraluminal surgery. The research 5)_____ that flexible robots enhance surgical flexibility and precision by increasing the range of motion and 6)_____ modular design. Additionally, the article emphasizes the spatial steerability and surgical precision of flexible robots, highlighting their 7)_____ with safety and 8)_____. Overall, flexible robots hold great potential in improving surgical outcomes and enhancing patients' quality of life in minimally invasive procedures. The article proposes several core principles, including spatial control, modular structuring, surgical precision, and 9)_____ ability, to guide the development of flexible robotic systems. Furthermore, the application of intelligent 10)_____ control models is considered crucial for ensuring safe and efficient intrabody navigation.

III. Group work: Work in groups and complete the following analysis of the structure with the information from Text B.

Function	Paragraph	Content
Introduction	A	Introduce the importance and 1)_____ of flexible robots in minimally invasive surgery (MIS); Present the main topic and purpose of the article.
Body	B	Describe the 2)_____ of flexible robots in visualizing and 3)_____ hard-to-reach anatomical organs; Mention the research and development of design and control systems for 4)_____ robotic surgery; Highlight the limited number of useful prototypes for procedures performed via natural or single incised port.
	C	Discuss the importance of spatial flexibility and modular design in 5)_____ surgical flexibility and precision; Emphasize the 6)_____ of chain-link structures and short interspaced actuations on robot flexibility and maneuverability.

Continued Table

Function	Paragraph	Content
Body	D	Highlight the importance of surgical precision in robotic surgery; Mention the need for intelligent 7) _____ control models for precise intrabody navigation.
	E	Explore the 8) _____ of intelligent constraint control models in flexible robot navigation; Stress the role of these models in achieving accurate and timely navigation.
Conclusion	F	Summarize the 9) _____ and future directions of flexible robots in MIS; Highlight the 10) _____ of improving surgical precision and manipulation capabilities with flexible robots.

Academic Expression

Pair work: Discuss with your partner and compare the possible paraphrases of each sentence and decide which one expresses the original meaning more academically and accurately.

1. Flexible robots enable surgeons to visualize and manipulate anatomical organs that are difficult to access with previous robots used in MIS.

   a. Flexible robots let surgeons see and handle organs that were tough to get to with the old robots used in minimally invasive surgery.

   b. Flexible robots facilitate the visualization and manipulation of anatomical organs that were challenging to reach with earlier generations of robotic systems utilized in minimally invasive surgery.

2. Several studies have been dedicated to the design and development of control systems for flexible robotic surgery.

   a. Extensive research efforts have been devoted to the design and development of control systems specifically tailored for flexible robotic surgery.

   b. Many studies have focused on designing and developing control systems for flexible robotic surgery.

3. Only a few prototypes are found useful for procedures performed via natural or single incised port.

    a. Only a few prototypes have proven to be useful for procedures done through a natural or single incised port.

    b. A limited number of prototypes have demonstrated utility for procedures conducted through a natural or single incised port.

4. Spatially flexible robots can be utilized in delivering large dosage of morphological and functional data for the treatment of cancerous organs.

    a. Spatially flexible robots have the potential to deliver substantial amounts of morphological and functional data for the treatment of cancerous organs.

    b. Flexible robots that can move in different directions can provide a lot of information about the shape and function of cancerous organs.

5. Omisore et al. proposed a flexible robotic system for minimally invasive radiosurgery of gastrointestinal tumors.

    a. Omisore and colleagues came up with a flexible robot system for doing less invasive radiosurgery on gastrointestinal tumors.

    b. Omisore et al. put forward a flexible robotic system designed for performing minimally invasive radiosurgery on gastrointestinal tumors.

6. The utilization of modular structuring becomes a fundamental principle for ensuring the suitability of flexible robots in single-port intraluminal surgery.

    a. The adoption of modular structuring serves as a fundamental principle to guarantee the applicability of flexible robots in single-port intraluminal surgery.

    b. Using modules is a really important rule to make sure that flexible robots work well in surgeries where they only make one small cut.

Understanding the Text

Ⅰ. Text B has six paragraphs, A-F. Choose the correct heading for paragraphs A-F from the list of headings below.

List of headings

Ⅰ	Surgical Precision in Robotic Surgery
Ⅱ	Single-Port Access Surgery

Continued Table

III	Benefits of Flexible Robotic Systems in Surgical Procedures
IV	Spatial Steerability of Flexible Robots
V	Flexible Robots for Minimally Invasive Surgery
VI	Core Principles for Safe and Efficient Robotic Manipulation
VII	Range of Motion in Flexible Surgical Robots
VIII	Challenges and Limitations of Flexible Robotic Surgery

Paragraph A:
Paragraph B:
Paragraph C:
Paragraph D:
Paragraph E:
Paragraph F:

II. Do the following statements agree with the information given in Text B? Write your judgement.

TRUE	if the statement agrees with the information
FALSE	if the statement contradicts the information
NOT GIVEN	if there is no information on this

1. Flexible robots are primarily used for cosmetic surgeries and have limited applications in other medical procedures.

2. Some prototypes are found useful for procedures performed via natural or single incised port.

3. Spatially flexible robots can be utilized in delivering large dosage of morphological and functional data for the treatment of cancerous organs.

4. Omisore et al. proposed a flexible robotic system for minimally invasive radiosurgery of gastrointestinal tumors.

5. Single-port access surgery enables surgeons to perform complex procedures through spatially flexible and narrow access.

6. Distally actuated devices are utilized to avoid collisions in the surgical workspace.

7. Modular underactuation is a disadvantage of the design principle for flexible surgical robots.

8. Surgical precision is of great importance in robotic surgery, and flexible robotics allows for better surgical tasks with geometric accuracy and precision.

9. Spatially flexible robots have shorter range of motion compared to conventional arm robots.

10. Modular underactuation allows for greater flexibility and precision in the movements of flexible surgical robots.

Ⅲ. **Text B has six paragraphs, A-F. Which paragraph contains the following information? You may use any letter more than once.**

1. Flexible surgical robots possess enhanced range of motion due to a series of short interspaced actuations in their chain of serial links.

2. Modular structuring is a fundamental principle for enabling flexible robots to be suitable for single-port intraluminal surgery.

3. Single-port access represents a groundbreaking advancement in intra-body robotic surgery.

4. Flexible robotics has been previously acclaimed in literature as a superior method for achieving geometric accuracy and precision in surgical tasks.

5. Only a handful of prototypes have proven useful for procedures performed through natural or single incised ports.

6. Surgeons can guide the instrument through flexible pathways using a single port.

7. Flexible endoscopic robots employ high-definition cameras with magnifying capabilities for improved visual perception during surgery.

8. Previous robots used in MIS were unable to access anatomical organs effectively, but flexible robots now allow surgeons to visualize and manipulate them.

Rise of a Great Power

Translate the following Chinese part into English and make it a complete English text.

In recent years, China has witnessed rapid growth in the research, development, and application of medical robots, providing patients with safer and more precise medical services. 1)(外科机器人能够协助医生进行复杂和精细的手术,提高手术的准确性和成功率)____

_____

_____.

Rehabilitation robots help patients with rehabilitation training, enhancing their mobility and quality of life, including lower limb rehabilitation robots, upper limb rehabilitation robots, and stroke rehabilitation robots.

Universities, research institutes, and medical institutions in China 2)(设立了专门的医疗机器人实验室,培养了一批优秀的研究人员和工程师) _____

_____. Chinese medical robot companies have also gained a competitive advantage in both domestic and international markets, 3)(推动了医疗机器人技术的商业化和应用) _____

_____.

Despite the remarkable progress made in the field of medical robots in China, there are still challenges to overcome, 4)(制定和执行技术标准和法规是其中之一) _____

_____. Additionally, the high cost of medical robots remains a constraining factor. However, with the continuous advancement of technology and government support, 5)(中国有望在医疗机器人技术方面取得更大的突破,为人类健康事业做出更大的贡献) _____

_____.

4-1　Unit Four-Answer and Translation

Unit Five　Materials Science

Warm-up

The world is currently experiencing a profound scientific and technological revolution, accompanied by industrial transformation and remarkable breakthroughs. This revolution is characterized by the seamless fusion of advanced information technology and manufacturing, as well as groundbreaking developments in areas such as new energy, materials, and biotechnology. These advancements are driving extensive changes across various industries, leading to the emergence of innovative solutions, increased productivity, and new avenues for economic growth. This global transformation is reshaping the way we live, work, and interact, opening up endless possibilities for progress and shaping the future of our society.

Thought-provoking Questions

· In the context of the profound scientific and technological revolution and industrial transformation, how can mechanical engineering professionals contribute to the development and implementation of innovative solutions that harness the fusion of advanced information technology and manufacturing?

· How can mechanical engineering students prepare themselves to thrive in an era of remarkable breakthroughs and extensive changes driven by the scientific and technological revolution?

Text A
Materials Science: Interdisciplinary Research and Applications

Research Background

Materials science is an interdisciplinary field that studies materials, while materials engineering focuses on designing and improving materials for various applications. It originated from the Age of Enlightenment, combining analytical thinking from chemistry, physics, and engineering. Over time, it evolved into a distinct field, with dedicated schools and research areas. Understanding the relationship between processing, structure, and properties is crucial in materials science. It has contributed to advancements in nanotechnology, biomaterials, and metallurgy. The field has expanded to include a wide range of materials, and computer simulations are now used to discover new materials and predict their properties.

Introduction Questions

· How did the intellectual origins of materials science contribute to its development as a distinct field?

· What is the significance of the materials paradigm in materials science research?

· How has the field of materials science evolved over time, and what factors contributed to its growth?

A. Materials science is an interdisciplinary field of researching and discovering materials. Materials engineering is an engineering field of designing and improving materials, and finding uses for materials in other fields and industries.

B. The intellectual origins of materials science stem from the Age of Enlightenment, when researchers began to use analytical thinking from chemistry, physics, and engineering to understand ancient, phenomenological observations in metallurgy and mineralogy.

C. Materials science still incorporates elements of physics, chemistry, and engineering. As such, the field was long considered by academic institutions as a sub-field of these related fields. Beginning in the 1940s, materials science began to be more widely recognized as a specific and distinct field of science and engineering, and major technical universities around the world created dedicated schools for its study.

D. Materials scientists emphasize understanding how the history of a material (processing) influences its structure, and thus the material's properties and performance. The understanding of processing-structure-properties relationships is called the materials paradigm. This paradigm is used to advance understanding in a variety of research areas, including nanotechnology, biomaterials, and metallurgy.

E. Materials science is also an important part of forensic engineering and failure analysis-investigating materials, products, structures or components, which fail or do not function as intended, causing personal injury or damage to property. Such investigations are key to understanding, for example, the causes of various aviation accidents and incidents.

F. The material of choice of a given era is often a defining point. Phrases such as Stone Age, Bronze Age, Iron Age, and Steel Age are historic, if arbitrary examples. Originally deriving from the manufacture of ceramics and its putative derivative metallurgy, materials science is one of the oldest forms of engineering and applied science. Modern materials science evolved directly from metallurgy, which itself evolved from the use of fire. A major breakthrough in the understanding of materials occurred in the late 19th century, when the American scientist Josiah Willard Gibbs demonstrated that the thermodynamic properties related to atomic structure in various phases are related to the physical properties of a material. Important elements of modern materials science were products of the Space Race; the understanding and engineering of the metallic alloys, and silica and carbon materials, used in building space vehicles enabling the exploration of space. Materials science has driven, and been driven by, the development of revolutionary technologies such as rubbers, plastics, semiconductors, and biomaterials.

G. Before the 1960s (and in some cases decades after), many eventual materials science departments were metallurgy or ceramics engineering departments, reflecting the 19th and early 20th century emphasis on metals and ceramics. The growth of materials science in the United States was catalyzed in part by the Advanced Research Projects Agency, which funded a series of university-hosted laboratories in the early 1960s, "to expand the national program of basic research and training in the materials sciences." In comparison with mechanical engineering, the nascent material science field focused on addressing materials from the macro-level and on the approach that materials are designed on the basis of knowledge of behavior at the microscopic level. Due to the expanded knowledge of the link between atomic and molecular processes as well as the overall properties of materials, the design of materials came to be based on specific desired properties. The materials science field has since broadened to include every class of materials, including ceramics, polymers, semiconductors, magnetic materials, biomaterials, and nanomaterials, generally classified into three distinct groups: ceramics, metals, and polymers. The prominent change in materials science during the recent decades is active usage of computer simulations to find new materials, predict properties and understand phenomena.

H. A material is defined as a substance (most often a solid, but other condensed phases can be included) that is intended to be used for certain applications. There are a myriad of materials around us; they can be found in anything from buildings and cars to spacecraft. The main classes of materials are metals, semiconductors, ceramics and polymers. New and advanced materials that are being developed include nanomaterials, biomaterials, and energy materials to name a few.

I. The basis of materials science is studying the interplay between the structure of materials, the processing methods to make that material, and the resulting material properties. The complex combination of these produce the performance of a material in a specific application. Many features across many length scales impact material performance, from the constituent chemical elements, its microstructure, and macroscopic features from processing. Together with the laws of thermodynamics and kinetics materials scientists aim to understand and improve materials.

Vocabulary List

phenomenological
relating to the study of conscious experience and phenomena as perceived 现象学的

metallurgy
the branch of science and technology concerned with the properties of metals and their production and purification 冶金学

mineralogy
the scientific study of minerals 矿物学

nanotechnology

the branch of technology that deals with dimensions and tolerances of less than 100 nanometers, especially the manipulation of individual atoms and molecules   纳米技术

biomaterials

materials that are compatible with biological systems and can be used in medical applications   生物材料

putative

generally considered or reputed to be   推测的

metallic alloys

mixtures composed of two or more metallic elements   金属合金

silica

a hard, unreactive, colorless compound that occurs as the mineral quartz and as a principal constituent of sandstone and other rocks   二氧化硅

semiconductors

materials that have electrical conductivity intermediate between that of metals and insulators and can be controlled by adding impurities or applying electric fields   半导体

polymers

substances or materials consisting of large molecules made up of repeating structural units   聚合物

kinetics

the branch of chemistry or biochemistry concerned with the rate of chemical reactions   动力学

Language Enhancement

Ⅰ. Complete the following sentences with words listed in the box below. Change the form where necessary.

scale	interplay	condense	prominent	broaden
impact	catalyze	nascent	putative	derivative

Continued Table

scale	interplay	condense	prominent	broaden
constituent	stem	analytical	property	aviation

1. The development of nanotechnology has significantly broadened the _____ of possibilities in various scientific fields.

2. In the _____ between different technologies, artificial intelligence and robotics have had a prominent impact on automation.

3. To improve data storage efficiency, scientists are working on _____ large amounts of information into smaller devices.

4. Elon Musk is a _____ figure in the aerospace industry due to his significant contributions to the field of aviation and space exploration.

5. The advancement of renewable energy technologies has _____ our understanding of the constituent elements required for sustainable power generation.

6. The _____ of technological innovation lies in continuous research and development efforts.

7. The _____ approach in data science enables us to extract valuable insights from vast amounts of information.

8. Cybersecurity is a critical _____ of modern technology, as it ensures the protection of sensitive data and prevents unauthorized access.

9. The advent of electric _____ has the potential to catalyze a major transformation in the transportation industry.

10. The nascent field of quantum computing holds _____ promises for solving complex computational problems efficiently.

11. Machine learning is a _____ technology that has emerged from the broader field of artificial intelligence.

12. User experience is a _____ aspect of software design, aiming to create intuitive and user-friendly interfaces.

13. The discovery of antibiotics had a profound _____ on modern medicine, revolutionizing the treatment of infectious diseases.

14. The _____ field of quantum computing holds great promise for solving complex computational problems.

15. The government's funding initiatives aim to _____ the development of nascent industries.

Ⅱ. **Complete the summary using the list of words in the box. Change the form where necessary.**

interplay	dedicate	origin	diverse	aviation
analytical	foster	metallurgy	broaden	enhance
application	significant	mineralogy	property	paradigm

Materials science is an interdisciplinary field that investigates the 1)_____, structure, and 2)_____ of materials. Its 3)_____ can be traced back to the Age of Enlightenment, where 4)_____ thinking from chemistry, physics, and engineering was applied to understand 5)_____ and 6)_____. Over time, materials science gained recognition as a distinct field, leading to the establishment of 7)_____ schools worldwide.

The field focuses on the 8)_____ between processing, structure, and properties of materials, known as the materials 9)_____. It has 10)_____ applications, including nanotechnology, biomaterials, and forensic engineering. Materials science has played a 11)_____ role in understanding and improving material performance, leading to advancements in industries such as 12)_____.

Historically, materials have defined different ages, such as the Stone Age and Bronze Age. Breakthroughs in the late 19th century, particularly in thermodynamics, 13)_____ our understanding of materials at the atomic level. The field has evolved, 14)_____ its scope to include ceramics, polymers, semiconductors, and nanomaterials.

Overall, materials science drives technological innovations, 15)_____ interdisciplinary research, and contributes to the understanding and development of materials for various applications.

Academic Expression

Pair work: Discuss with your partner and compare the two possible paraphrases of each sentence and decide which one expresses the original meaning more academically.

1. The intellectual origins of materials science stem from the Age of Enlightenment, when researchers began to use analytical thinking from chemistry, physics, and engineering to understand ancient, phenomenological observations in metallurgy and mineralogy.

   a. The roots of materials science can be traced back to the Age of Enlightenment, wherein scholars employed analytical methodologies derived from the disciplines of chemistry, physics, and engineering to comprehend the ancient and phenomenological phenomena associated with metallurgy and mineralogy.

　　b. The beginnings of materials science can be traced back to the Age of Enlightenment when researchers started using analytical thinking from chemistry, physics, and engineering to understand ancient observations in metallurgy and mineralogy.

2. Materials science still incorporates elements of physics, chemistry, and engineering. As such, the field was long considered by academic institutions as a sub-field of these related fields.

　　a. Materials science continues to integrate principles from physics, chemistry, and engineering. Consequently, academic institutions have historically regarded it as a sub-discipline within these interconnected fields.

　　b. Materials science still combines aspects of physics, chemistry, and engineering. That's why academic institutions have long seen it as a subset of these related fields.

3. Materials scientists emphasize understanding how the history of a material (processing) influences its structure, and thus the material's properties and performance.

　　a. Material scientists place significant emphasis on comprehending how the processing history of a material impacts its structure, subsequently influencing its properties and overall performance.

　　b. Materials scientists focus on understanding how the way a material is processed affects its structure, and ultimately, its properties and performance.

4. The prominent change in materials science during the recent decades is active usage of computer simulations to find new materials, predict properties and understand phenomena.

　　a. A notable development in materials science in recent decades involves the extensive utilization of computer simulations for the discovery of novel materials, property prediction, and the exploration of various phenomena.

　　b. One significant advancement in materials science over the past few decades is the widespread use of computer simulations to discover new materials, predict properties, and gain insights into different phenomena.

5. The materials science field has since broadened to include every class of materials, including ceramics, polymers, semiconductors, magnetic materials, biomaterials, and nanomaterials, generally classified into three distinct groups: ceramics, metals, and polymers.

a. The scope of materials science has expanded significantly to encompass all classes of materials, including ceramics, polymers, semiconductors, magnetic materials, biomaterials, and nanomaterials, which are broadly categorized into three distinct groups: ceramics, metals, and polymers.

b. The field of materials science has grown to include a wide range of materials, such as ceramics, polymers, semiconductors, magnetic materials, biomaterials, and nanomaterials. These materials are generally divided into three main groups: ceramics, metals, and polymers.

6. The understanding of processing-structure-properties relationships is called the materials paradigm.

a. The comprehension of the interrelationships between processing, structure, and properties is referred to as the materials paradigm.

b. The materials paradigm refers to the understanding of how processing, structure, and properties are interconnected.

7. Materials science is an interdisciplinary field that combines principles from physics, chemistry, and engineering to explore the structure, properties, and applications of materials.

a. Materials science represents an interdisciplinary domain that integrates principles from physics, chemistry, and engineering to investigate the structure, properties, and applications of diverse materials.

b. Materials science is a multidisciplinary field that brings together concepts from physics, chemistry, and engineering to study the structure, properties, and uses of materials.

8. Overall, materials science drives technological innovations, fosters interdisciplinary research, and contributes to the understanding and development of materials for various applications.

a. In essence, materials science propels technological advancements, encourages interdisciplinary investigations, and significantly contributes to the comprehension and advancement of materials for a multitude of applications.

b. In summary, materials science is a driving force behind technological advancements, promotes collaboration across disciplines, and plays a crucial role in understanding and developing materials for various practical uses.

Understanding the Text

Ⅰ. **Pair work: Work with your partner and take turns asking and answering the following questions according to the information contained in the text.**

1. What were the intellectual origins of materials science?

2. When did materials science begin to be recognized as a specific field of science and engineering?

3. What is the materials paradigm?

4. How does materials science contribute to forensic engineering and failure analysis?

5. What are some historic examples of material-based ages?

6. How did the understanding of materials evolve during the late 19th century?

7. What catalyzed the growth of materials science in the United States?

8. How has the materials science field broadened over time?

Ⅱ. **Group work: Please work in groups and complete the structural table with the information from Text A.**

Section	Paragraph	Content
Introduction	A	Definition and 1) _____ between materials science and materials engineering.
Main Body	B	The intellectual origins of materials science from the Age of Enlightenment and its 2) _____.
	C	Recognition of materials science as a distinct field and the 3) _____ of dedicated schools.

Continued Table

Section	Paragraph	Content
Main Body	D	Emphasis on understanding the processing-structure-properties relationships and the materials 4)_____.
	E	The importance of materials science in forensic engineering and failure analysis.
Historical Development	F	Historical context of materials science, including the influence of different ages and 5)_____ in understanding materials.
	G	Growth and 6)_____ of materials science as a field, including the role of the Advanced Research Projects Agency and the 7)_____ scope of materials.
Key Concepts and Applications:	H	8)_____ of materials and the introduction of new and advanced materials.
Conclusion	I	9)_____ between the structure, processing methods, and resulting properties of materials, with a focus on understanding and improving material performance.
		Recap of the interdisciplinary nature and 10)_____ of materials science.

Rise of a Great Power

Translate the following Chinese part into English and make it a complete English text.

Against the backdrop of rapid economic growth and technological innovation, 1)(中国的材料科学正迅速崛起,并成为全球材料研究的重要力量。)_____

_____2)(中国的材料科学研究广泛应用于许多领域)_____

_____, such as electronics, optoelectronics, energy, healthcare, and aerospace. Through the development of new materials, 3)(中国科研人员取得了一系列重要突破)_____

_____. For instance, the development of novel semiconductor materials has opened up new possibilities for the electronics industry, 4)(高效能源材料的研究推动了清洁能源的发展,生物材料的创新提高了医疗技术和治疗效果。)_____

_____.

Chinese researchers will continue their efforts to promote research and innovation in materials science, 5)(推动材料科学的研究与创新,为实现经济转型升级、推动可持续发展做出积极贡献。)_____

_____.

Text B
Material Structure: From Atoms to Macroscopic Scale

Research Background

Materials science is dedicated to studying the relationship between the structures and properties of materials. This encompasses the examination of materials at various scales, from atomic to macroscopic levels. Characterization techniques, such as diffraction, spectroscopy, and chemical analysis, are employed to explore material structure. The atomic structure focuses on arrangements and bonding, influencing electrical, magnetic, and chemical properties. Crystallography investigates crystalline arrangements and defects, crucial for understanding material behavior. Nanostructure analysis uncovers unique properties in the nanoscale range. Microstructure analysis reveals how arrangements at the micrometer scale affect physical properties. Macrostructure investigation provides a visual understanding of materials at larger scales.

> **Introduction Questions**
> · How does the atomic structure influence the properties of materials?
> · What role does crystallography play in understanding material properties?
> · How do nanostructures contribute to the unique properties of materials?

Structure is one of the most important components of the field of materials science. The very definition of the field holds that it is concerned with the investigation of "the relationships that exist between the structures and properties of materials". Materials science examines the structure of materials from the atomic scale, all the way up to the macro scale. Characterization is the way materials scientists examine the structure of a material. This involves methods such as diffraction with X-rays, electrons or neutrons, and various forms of spectroscopy and chemical analysis such as Raman spectroscopy, energy dispersive spectroscopy, chromatography, thermal analysis, electron microscope analysis, etc.

A. Atomic structure

Atomic structure deals with the atoms of the materials, and how they are arranged to give rise to molecules, crystals, etc. Much of the electrical, magnetic and chemical properties of materials arise from this level of structure. The length scales involved are in angstroms. The chemical bonding and atomic arrangement (crystallography) are fundamental to studying the properties and behavior of any material.

B. Bonding

To obtain a full understanding of the material structure and how it relates to its properties, the materials scientist must study how the different atoms, ions and molecules are arranged and bonded to each other. This involves the study and use of quantum chemistry or quantum physics. Solid-state physics, solid-state chemistry and physical chemistry are also involved in the study of bonding and structure.

C. Crystallography

Crystallography is the science that examines the arrangement of atoms in crystalline solids. Crystallography is a useful tool for materials scientists. In single crystals, the effects of the crystalline arrangement of atoms is often easy to see macroscopically, because the natural shapes of crystals reflect the atomic structure. Further, physical properties are often controlled by crystalline defects. The understanding of crystal structures is an important prerequisite for understanding crystallographic defects. Mostly, materials do not occur as a single crystal, but in polycrystalline form, as an aggregate of small crystals or grains with different orientations. Because of this, the powder diffraction method, which uses diffraction patterns of polycrystalline samples with a large number of crystals, plays an important role in structural determination. Most materials have a crystalline structure, but some important materials do not exhibit regular crystal structure.

Polymers display varying degrees of crystallinity, and many are completely non-crystalline. Glass, some ceramics, and many natural materials are amorphous, not possessing any long-range order in their atomic arrangements. The study of polymers combines elements of chemical and statistical thermodynamics to give thermodynamic and mechanical descriptions of physical properties.

D. Nanostructure

Materials, which atoms and molecules form constituents in the nanoscale (i. e., they form nanostructure) are called nanomaterials. Nanomaterials are subject of intense research in the materials science community due to the unique properties that they exhibit.

Nanostructure deals with objects and structures that are in the 1-100 nm range. In many materials, atoms or molecules agglomerate together to form objects at the nanoscale. This causes many interesting electrical, magnetic, optical, and mechanical properties. In describing nanostructures, it is necessary to differentiate between the number of dimensions on the nanoscale. Nanotextured surfaces have one dimension on the nanoscale, i. e., only the thickness of the surface of an object is between 0.1 and 100 nm. Nanotubes have two dimensions on the nanoscale, i. e., the diameter of the tube is between 0.1 and 100 nm; its length could be much greater.

Finally, sphericalnanoparticles have three dimensions on the nanoscale, i. e., the particle is between 0.1 and 100 nm in each spatial dimension. The terms nanoparticles and ultrafine particles (UFP) often are used synonymously although UFP can reach into the micrometre range. The term "nanostructure" is often used, when referring to magnetic technology. Nanoscale structure in biology is often called ultrastructure.

E. Microstructure

Microstructure is defined as the structure of a prepared surface or thin foil of material as revealed by a microscope above 25× magnification. It deals with objects from 100 nm to a few cm. The microstructure of a material (which can be broadly classified into metallic, polymeric, ceramic and composite) can strongly influence physical properties such as strength, toughness, ductility, hardness, corrosion resistance, high/low temperature behavior, wear resistance, and so on. Most of the traditional materials (such as metals and ceramics) are microstructured.

The manufacture of a perfect crystal of a material is physically impossible. For example, any crystalline material will contain defects such as precipitates, grain boundaries (Hall-Petch relationship), vacancies, interstitial atoms or substitutional atoms. The microstructure of materials reveals these larger defects and advances in simulation have allowed an increased understanding of how defects can be used to enhance material properties.

F. Macrostructure

Macrostructure is the appearance of a material in the scale millimeters to meters, it is the structure of the material as seen with the naked eye.

Vocabulary List

diffraction
the process by which a beam of light or other waves spreads 衍射

electrons
subatomic particles with a negative charge 电子

neutrons
subatomic particles with no charge 中子

spectroscopy
the study of the interaction between matter and electromagnetic radiation 光谱学

chromatography
a technique used to separate and analyze mixtures of compounds 色谱法

crystallography
the science of determining the arrangement of atoms in crystals 结晶学

crystalline
having a regular, repeating arrangement of atoms 结晶的

ultrafine particles (UFP)
extremely small particles, often referring to particles less than 100 nanometers in size 超细颗粒(UFP)

corrosion

the gradual deterioration of a material due to chemical reactions 腐蚀

precipitates

solid particles that form from a solution during a chemical reaction or phase change 沉淀物

amorphous

without a defined or regular shape or structure 无定形的

agglomerate

to collect or form into a mass or cluster 凝聚

ductility

the ability of a material to stretch or deform without breaking 延展性

Language Enhancement

I. Choose the most suitable word in the box below to complete each sentence. Change the form where necessary.

diffraction	dimension	synonymous	resistance	orientation
differentiate	interstitial	magnification	corrosion	aggregate

1. _____ doping is commonly used in semiconductor fabrication to introduce impurities into the crystal lattice, altering its electrical properties.

2. _____-resistant coatings are applied to metal surfaces to protect them from degradation caused by exposure to harsh environments.

3. In the field of artificial intelligence, the terms "machine learning" and "deep learning" are often used _____, although they have distinct meanings.

4. To _____ between two similar DNA sequences, scientists use advanced sequencing techniques that can identify subtle variations in nucleotide sequences.

5. _____ data from multiple sources allows researchers to analyze large datasets and extract meaningful insights.

6. The _____ of a circuit component determines how much current it allows to flow under a given voltage, affecting the overall performance of the electronic system.

7. X-ray _____ is a powerful technique used in materials science to study the atomic structure and crystallography of various materials.

8. High _____ microscopy techniques enable scientists to observe nanoscale structures and analyze their properties with great detail.

9. The _____ of a dataset refers to the number of variables or features used to represent each data point, impacting the complexity of data analysis algorithms.

10. The _____ of magnetic domains in a material affects its magnetic properties, making it suitable for applications such as data storage in hard drives.

Ⅱ. **Complete the summary using the suitable words in the Text B.**

Materials science is an 1) _____ field focused on the study and application of various materials. Originating from the Age of 2) _____, it combines 3) _____ thinking from chemistry, physics, and engineering to understand metallurgical and mineralogical phenomena. In the 1940s, materials science gained 4) _____ as a distinct discipline, leading to the establishment of schools worldwide.

This field encompasses a wide range of materials, including ceramics, polymers, and 5) _____. It plays a 6) _____ role in technological advancements, such as nanotechnology and 7) _____, and is essential in forensic engineering and failure 8) _____. Crystallography aids in understanding atomic structures and their 9) _____ on material properties.

Materials science continuously 10) _____ the relationship between structure, processing, and properties, 11) _____ the design and improvement of materials for specific applications. Computer simulations have revolutionized the 12) _____ of new materials and prediction of their 13) _____. From the Stone Age to the present, materials have driven human 14) _____, and materials science remains at the forefront of research and 15) _____, shaping the technologies of the future.

Academic Expression

Pair work: Discuss with your partner and compare the two possible paraphrases of each sentence and decide which one expresses the original meaning more academically.

1. The understanding of crystal structures is an important prerequisite for understanding crystallographic defects.

   a. You gotta understand crystal structures before you can fully get crystallographic defects.

b. A comprehensive comprehension of crystal structures is a crucial requirement for grasping crystallographic defects.

2. Nanomaterials are subject of intense research in the materials science community due to the unique properties that they exhibit.

    a. Nanomaterials are the focal point of extensive research in the materials science community owing to their distinct properties.

    b. Materials scientists are going all out in researching nanomaterials because they have some really special properties.

3. The microstructure of a material can strongly influence physical properties such as strength, toughness, ductility, hardness, corrosion resistance, high/low temperature behavior, wear resistance, and so on.

    a. The way a material is structured can totally affect its physical properties, like how strong it is, how well it resists damage, how flexible it is, how hard it is, how well it fights corrosion, and even how it behaves at different temperatures.

    b. The microstructure of a material exerts a significant impact on its physical properties, including strength, toughness, ductility, hardness, corrosion resistance, high/low temperature behavior, wear resistance, and other factors.

4. Most of the traditional materials (such as metals and ceramics) are microstructured.

    a. The majority of conventional materials, such as metals and ceramics, possess a microstructure.

    b. Traditional materials, like metals and ceramics, have a microstructure.

5. Macrostructure is the appearance of a material in the scale millimeters to meters, it is the structure of the material as seen with the naked eye.

    a. Macrostructure is basically how a material looks to the naked eye, like its overall appearance in terms of millimeters to meters.

    b. Macrostructure refers to the visible features of a material at the scale of millimeters to meters, representing its structural characteristics discernible to the naked eye.

6. The study of polymers combines elements of chemical and statistical thermodynamics to give thermodynamic and mechanical descriptions of physical properties.

    a. The investigation of polymers integrates principles from chemical and statistical thermodynamics to provide thermodynamic and mechanical explanations for their physical properties.

b. When we study polymers, we mix chemistry and thermodynamics to explain their physical properties.

7. Crystallography is a useful tool for materials scientists.

a. Crystallography comes in handy for materials scientists.

b. Crystallography serves as a valuable tool for materials scientists.

8. In many materials, atoms or molecules agglomerate together to form objects at the nanoscale, causing many interesting electrical, magnetic, optical, and mechanical properties.

a. In numerous materials, the aggregation of atoms or molecules results in the formation of nanostructures, contributing to a wide range of intriguing electrical, magnetic, optical, and mechanical properties.

b. In lots of materials, atoms or molecules get all clumped up together at the nanoscale, and that's what gives them those cool electrical, magnetic, optical, and mechanical properties.

 Understanding the Text

Ⅰ. **Text B has six paragraphs, A-F. Which paragraph contains the following information? You may use any letter more than once.**

1. The field of materials science gained recognition as a distinct discipline in the 1940s.

2. In the late 19th century, Josiah Willard Gibbs achieved a major breakthrough in the understanding of materials.

3. Materials science covers various types of materials, such as ceramics, polymers, and semiconductors.

4. The Advanced Research Projects Agency played a pivotal role in fostering the growth of materials science in the United States.

5. Forensic engineering and failure analysis rely heavily on the contributions of materials science.

6. Materials science has expanded its scope to encompass nanomaterials and biomaterials.

7. Grasping the relationship between processing, structure, and properties is fundamental to materials science.

8. Technological advancements and the innovation of novel materials drive the progress of materials science.

9. Crystallography plays a vital role in comprehending the atomic arrangement of materials.

10. Materials science encompasses perspectives at both the macroscopic and microscopic levels.

II. Do the following statements agree with the information given in Text B? Write your judgement.

TRUE	if the statement agrees with the information
FALSE	if the statement contradicts the information
NOT GIVEN	if there is no information on this

1. Atomic structure deals with the arrangement of atoms in materials, and it has a significant impact on the properties of materials.
2. Crystallography is only useful for the study of single crystals and has limited relevance for polycrystalline materials.
3. Nanotubes have one dimension on the nanoscale, while spherical nanoparticles have three dimensions on the nanoscale.
4. The microstructure of a material can significantly affect its physical properties.
5. The manufacture of a perfect crystal of a material is physically impossible because all crystalline materials contain defects.
6. Macrostructure refers to the appearance of a material as seen with a microscope.
7. Materials scientists use diffraction with X-rays, electrons, or neutrons to examine the structure of materials.
8. Polymers are always completely non-crystalline and do not possess any long-range order in their atomic arrangements.
9. Electron microscopy is commonly used in materials science to investigate the microstructure of materials at high magnification.

10. Materials science also involves the study of materials' mechanical properties, such as elasticity, strength, and fracture toughness.

Ⅲ. **Group work**: **Please work in groups and complete the structural table with the information from Text B.**

Section	Description
Introduction	Introduces the importance of structure in materials science and its relationship with material 1)_____.
A. Atomic structure	Discusses the 2)_____ of atoms in materials and its impact on properties.
B. Bonding	Focuses on the study of how atoms, ions, and 3)_____ are arranged and bonded to each other.
C. Crystallography	4)_____ the examination of atomic arrangement in crystalline solids and its 5)_____ in materials science.
D. Nanostructure	6)_____ materials at the nanoscale, including nanomaterials and their 7)_____ properties.
E. Microstructure	Discusses the structure of materials at the microscale and its 8)_____ on physical properties.
F. Macrostructure	9)_____ the appearance of materials at a 10)_____ scale, visible to the naked eye.

Rise of a Great Power

Translate the following Chinese text into English.

China has made significant breakthroughs in the field of materials science through long-term research investment and innovative efforts.

1)（大量的科研机构和高校致力于材料结构的探索,从原子、晶体到微观结构,涉及了多个层面的研究。）_____

_____. Through advanced experimental techniques and simulation methods, 2)(中国科学家们揭示了许多材料的结构与性能之间的关联,为材料设计和应用提供了重要的基础)_____

_____

_____.

  Furthermore, 3)(中国引进和自主研发了一系列高分辨率的表征设备)_____

_____, such as electron microscopes, X-ray diffraction, and neutron diffraction, which have enhanced the observation and analysis capabilities of material structures. 4)(这些先进的技术推动了材料研究的进步,促进了新材料的发现和应用。)_____

_____

_____.

  In addition, the study and application of nanomaterials have become a hot research area. Chinese research teams have actively explored fields such as nanotubes, nanoparticles, and nanosurface structures. 5)(这些纳米结构展现出了独特的性能和潜在的应用价值,对电子学、光学和能源等领域具有重要意义)_____

_____

_____.

5-1 Unit Five-Answer and Translation

Unit Six　Systems Engineering

Warm-up

Mission of a Great Power

Implementing the strategy of building a strong manufacturing country is in line with the fundamental requirements of the "two centenary goals" and the "Chinese Dream". To achieve these goals, there must be a solid economic foundation and a strong manufacturing industry to support them, which urgently requires China's manufacturing industry to transform and upgrade, better meet the needs of economic and social development and national defense construction. By investing in technological advancements, innovation, and a skilled workforce, China can enhance its manufacturing capabilities, strengthen its economic foundation, and contribute to its overall progress.

Thought-provoking Questions

· How can the transformation and upgrading of China's manufacturing industry contribute to achieving the "two centenary goals" and the "Chinese Dream"?

· What are the key factors and challenges that China's manufacturing industry needs to address in its transformation and upgrading process?

Text A
Holistic Approaches for Complex Systems

Research Background

Systems engineering has evolved from physical systems to a broader concept, encompassing enterprise and service systems. It manages complexity, bridging the gap between user requirements and technical specifications. Its application extends beyond defense and aerospace to industries like IT, software development, and electronics. Research indicates that systems engineering reduces costs and brings benefits. Techniques such as modeling, simulation, and safety engineering validate assumptions and support critical decision-making. Approaches like soft systems methodology, system dynamics, and the Unified Modeling Language (UML) aid in the systems engineering process.

Introduction Questions

· How has the scope of systems engineering evolved over time, and what controversies surround its definition?

· How does systems engineering foster interdisciplinary collaboration in system development?

· Why is systems engineering crucial for managing complexity in systems and projects?

The traditional scope of engineering embraces the conception, design, development, production and operation of physical systems. Systems engineering, as originally conceived, falls within this scope. "Systems engineering", in this sense of the term, refers to the building of engineering concepts.

Evolution to broader scope

A. The use of the term "systems engineer" has evolved over time to embrace a wider, more holistic concept of "systems" and of engineering processes. This evolution of the definition has been a subject of ongoing controversy, and the term continues to apply to both the narrower and broader scope.

B. Traditional systems engineering was seen as a branch of engineering in the classical sense, that is, as applied only to physical systems, such as spacecraft and aircraft. More recently, systems engineering has evolved to take on a broader meaning especially when humans were seen as an essential component of a system. Peter Checkland, for example, captures the broader meaning of systems engineering by stating that "engineering can be read in its general sense; you can engineer a meeting or a political agreement."

C. Consistent with the broader scope of systems engineering, the Systems Engineering Body of Knowledge (SEBoK) has defined three types of systems engineering:
(1) Product Systems Engineering (PSE) is the traditional systems engineering focused on the design of physical systems consisting of hardware and software.
(2) Enterprise Systems Engineering (ESE) pertains to the view of enterprises, that is, organizations or combinations of organizations, as systems.
(3) Service Systems Engineering (SSE) has to do with the engineering of service systems.
Checkland defines a service system as a system which is conceived as serving another system. Most civil infrastructure systems are service systems.

Interdisciplinary field

D. System development often requires contribution from diverse technical disciplines. By providing a systems (holistic) view of the development effort, systems engineering helps mold all the technical contributors into a unified team effort, forming a structured development process that proceeds from concept to production to operation and, in some cases, to termination and disposal. In an acquisition, the holistic integrative discipline combines contributions and balances trade-offs among cost, schedule, and performance while maintaining an acceptable level of risk covering the entire life cycle of the item.

E. This perspective is often replicated in educational programs, in that systems engineering courses are taught by faculty from other engineering departments, which helps create an interdisciplinary environment.

Managing complexity

F. The need for systems engineering arose with the increase in complexity of systems and projects, in turn exponentially increasing the possibility of component friction, and therefore the unreliability of the design. When speaking in this context, complexity incorporates not only engineering systems, but also the logical human organization of data. At the same time, a system can become more complex due to an increase in size as well as with an increase in the amount of data, variables, or the number of fields that are involved in the design. The International Space Station is an example of such a system.

G. The development of smarter control algorithms, microprocessor design, and analysis of environmental systems also come within the purview of systems engineering. Systems engineering encourages the use of tools and methods to better comprehend and manage complexity in systems.

H. Taking an interdisciplinary approach to engineering systems is inherently complex since the behavior of and interaction among system components is not always immediately well defined or understood. Defining and characterizing such systems and subsystems and the interactions among them is one of the goals of systems engineering. In doing so, the gap that exists between informal requirements from users, operators, marketing organizations, and technical specifications is successfully bridged.

Scope

I. One way to understand the motivation behind systems engineering is to see it as a method, or practice, to identify and improve common rules that exist within a wide variety of systems. Keeping this in mind, the principles of systems engineering—holism, emergent behavior, boundary, etc.—can be applied to any system, complex or otherwise, provided systems thinking is employed at all levels. Besides defense and aerospace, many information and technology based companies, software development firms, and industries in the field of electronics & communications require systems engineers as part of their team.

J. An analysis by the INCOSE Systems Engineering center of excellence (SECOE) indicates that optimal effort spent on systems engineering is about 15%-20% of the total project effort. At the same time, studies have shown that systems engineering essentially leads to reduction in costs among other benefits. However, no quantitative survey at a larger scale encompassing a wide variety of industries has been conducted until recently. Such studies are underway to determine the effectiveness and quantify the benefits of systems engineering.

K. Systems engineering encourages the use of modeling and simulation to validate assumptions or theories on systems and the interactions within them. Use of methods that allow early detection of possible failures, in safety engineering, are integrated into the design process. At the same time, decisions made at the beginning of a project whose consequences are not clearly understood can have enormous implications later in the life of a system, and it is the task of the modern systems engineer to explore these issues and make critical decisions. No method guarantees today's decisions will still be valid when a system goes into service years or decades after first conceived. However, there are techniques that support the process of systems engineering. Examples include soft systems methodology, Jay Wright Forrester's System dynamics method, and the Unified Modeling Language (UML) — all currently being explored, evaluated, and developed to support the engineering decision process.

Vocabulary List

microprocessor

an integrated circuit that contains the functions of a central processing unit of a computer 微处理器

modeling

the process of creating a representation or simulation of a system or process   模型建立；塑造

synergy

the interaction or cooperation of two or more elements to produce a combined effect greater than the sum of their individual effects   同作用；协同效应

infrastructure

the basic physical and organizational structures and facilities needed for the operation of a society or enterprise   基础设施

termination

the action of bringing something to an end or the end of something   终止；结束

integrative

combining or coordinating separate elements into a unified whole   整合的；综合的

purview

the range of operation, authority, control, or concern 范围;权限

 Language Enhancement

Ⅰ. Choose the most suitable word in the box below to complete each sentence. Change the form where necessary.

conceive	quantify	termination	comprehend	incorporate
methodology	validate	emergent	disposal	evaluate
simulation	embrace	replicate	specification	infrastructure

1. We need to _____ the new methodology before implementing it in our research project.

2. By adopting a systematic _____, we have constructed a novel scientific research framework that provides comprehensive and reliable guidance for exploring future technological advancements.

3. It's important to _____ the effectiveness of the latest technology through rigorous testing.

4. The _____ accurately predicted the behavior of the chemical reaction, saving time and resources in the lab.

5. We need to _____ the impact of the new software on productivity to justify its implementation.

6. The _____ technologies in the field of artificial intelligence have the potential to revolutionize various industries.

7. The _____ document clearly outlines the requirements for the software development project.

8. It takes time to _____ complex algorithms, but with practice, one can become proficient in coding.

9. The software should _____ user-friendly features to enhance the overall user experience.

10. We successfully _____ the experiment in a different laboratory, confirming the previous findings.

11. Proper _____ of electronic waste is essential to prevent environmental pollution.

12. The robust _____ of the data center ensures high-speed connectivity and data storage.

13. The _____ of the project was due to budget constraints and a shift in company priorities.

14. We need to _____ renewable energy sources to reduce our dependency on fossil fuels.

15. The scientists _____ a groundbreaking idea for a new vaccine that could potentially eradicate a widespread disease.

Ⅱ. Complete the summary using the list of words in the box. Change the form where necessary.

collaboration	quantify	specification	integrate	holistic
conducive	utilizing	principle	challenge	facilitate
comprehend	emergent	encompass	validating	termination
crucial	interdisciplinary	interaction	enterprise	yields

Systems engineering has evolved from its traditional focus on physical systems to embrace a wider, more 1) _____ concept. It now 2) _____ the design and development of not only physical systems but also 3) _____ and service systems. This 4) _____ field helps manage complexity by providing a holistic view of system development and 5) _____ diverse technical disciplines into a unified team effort.

The 6) _____ of systems engineering, such as holism and 7) _____ behavior, can be applied to any system, making it relevant across industries like defense, aerospace, information technology, software development, and electronics.

Systems engineering plays a 8) _____ role in bridging the gap between informal requirements and technical 9) _____, ensuring effective communication and understanding among stakeholders. By 10) _____ tools and methods, systems engineering helps 11) _____ and manage complexity in systems, 12) _____ better decision-making and cost reduction.

The use of modeling and simulation aids in 13) _____ assumptions and understanding system 14) _____, while safety engineering techniques contribute to the design process. Systems engineering emphasizes the importance of considering the entire life cycle of a system, from concept to production, operation, and potential 15) _____ or disposal.

Efforts are underway to 16) _____ the benefits of systems engineering, with studies indicating that dedicating a portion of project effort to systems engineering 17) _____ cost reductions and other advantages. This interdisciplinary approach fosters 18) _____ among engineering departments and creates an environment 19) _____ to innovative problem-solving.

Overall, systems engineering provides a comprehensive framework for understanding and improving the functioning of complex systems, addressing the 20) _____ posed by increasing complexity in various industries and domains.

Academic Expression

Pair work: Discuss with your partner and compare the two possible paraphrases of each sentence and decide which one expresses the original meaning more academically.

1. Systems engineering, as originally conceived, falls within this scope.

   a. Systems engineering, as initially conceptualized, lies within this domain.

   b. Systems engineering, as it was first thought of, fits into this category.

2. The use of the term "systems engineer" has evolved over time to embrace a wider, more holistic concept of "systems" and of engineering processes.

   a. The way we refer to a "systems engineer" has changed over time to include a broader and more holistic idea of "systems" and how engineering works.

   b. The terminology "systems engineer" has undergone a gradual evolution to encompass a broader and more comprehensive understanding of "systems" and engineering methodologies.

3. Systems engineering encourages the use of tools and methods to better comprehend and manage complexity in systems.

   a. Systems engineering encourages using tools and methods to better understand and handle complex systems.

   b. Systems engineering promotes the utilization of tools and methodologies to enhance comprehension and management of system complexity.

4. Defining and characterizing such systems and subsystems and the interactions among them is one of the goals of systems engineering.

   a. Defining and characterizing such systems, subsystems, and their interactions constitute fundamental objectives within systems engineering.

   b. One of the main goals of systems engineering is to define and describe these systems and subsystems, and how they interact with each other.

5. Systems engineering provides a holistic view to mold all the technical contributors into a unified team effort.

a. Systems engineering offers a comprehensive perspective to integrate all technical contributors into a cohesive team endeavor.

　　b. Systems engineering gives us a complete view to bring all the technical contributors together as a united team.

6. No method guarantees today's decisions will still be valid when a system goes into service years or decades after first conceived.

　　a. There is no method that guarantees that the decisions made today will still be relevant when a system is used many years later.

　　b. No methodology assures the longevity of today's decisions when a system is deployed years or decades after its initial conception.

7. Systems engineering addresses the need for managing complexity in systems and projects.

　　a. Systems engineering tackles the requirement of complexity management in systems and projects.

　　b. Systems engineering deals with the challenge of handling complexity in systems and projects.

8. The International Space Station is an example of such a system.

　　a. The International Space Station exemplifies such a system.

　　b. The International Space Station is a perfect example of such a system.

Understanding the Text

　　Ⅰ. **Pair work**: Work with your partner and take turns asking and answering the following questions according to the information contained in the text.

1. What is the broader scope of systems engineering compared to its traditional focus?

2. What are the three types of systems engineering defined by the Systems Engineering Body of Knowledge (SEBoK)?

3. How does systems engineering help manage complexity in system development?

4. Why is systems engineering considered an interdisciplinary field?

5. What are some examples of systems that have become more complex?

6. What is the role of systems engineering in bridging the gap between informal requirements and technical specifications?

7. What areas of focus fall under the purview of systems engineering?

8. What is the significance of systems engineering in decision-making for a system's life cycle?

Ⅱ. Group work: Please work in groups and complete the structural table with the information from Text A.

Section	Paragraph	Key points
Introduction		The traditional scope of engineering includes the conception, design, development, production, and operation of physical systems.
Evolution to broader scope	A, B	Use of the term "systems engineer" has evolved to include a wider, more 1)_____ concept of "systems" and engineering processes. 2)_____ surrounds the evolving definition, and the term applies to both narrower and 3)_____ scopes. Systems engineering has broadened to include human involvement and various types of systems.
Interdisciplinary field	C, D, E	System development requires contributions from 4)_____ technical disciplines. Systems engineering provides a holistic view to 5)_____ technical contributors into a unified team. Educational programs often teach systems engineering through interdisciplinary 6)_____.
Managing complexity	F, G, H	Systems engineering addresses the need for managing 7)_____ in systems and projects. Complexity arises from the increase in system size, data, variables, and the logical human organization of data. Tools and methods are used to 8)_____ and manage complexity.
Scope	I	Systems engineering helps 9)_____ and improve common rules in a wide variety of systems. Principles of systems engineering can be applied to any system when 10)_____ systems thinking. Systems engineering is 11)_____ in defense, aerospace, technology, software development, and electronics industries.

Continued Table

Section	Paragraph	Key points
Quantitative analysis	J	Optimal effort spent on systems engineering is 12)_____ to be around 15%-20% of the total project effort. Studies indicate that systems engineering leads to cost reduction and other benefits, but larger-scale surveys are still 13)_____.
Modeling and simulation	K	Systems engineering promotes the use of modeling and simulation to 14)_____ assumptions and understand system interactions. Early 15)_____ of failures and critical decision-making are integral to the design process. Various techniques and methods support the systems engineering decision process.

Rise of a Great Power

Translate the following Chinese part into English and make it a complete English text.

With the rapid development of the economy and continuous technological advancements, 1)(系统工程在提高效率、优化资源利用和解决复杂问题方面发挥着关键作用)_____. Systems engineering is widely applied in China across various sectors, 2)(涉及交通运输、电力能源、制造业、信息技术、城市规划等诸多领域)_____. Through the use of systems engineering methodologies and tools, 3)(中国在基础设施建设、交通运输网络优化、能源系统规划和智慧城市建设等方面取得了显著成就)_____.

In the future, as technology continues to innovate and demands evolve, 4)(中国的系统工程将面临新的挑战和机遇)_____. Chinese researchers and engineers will continue their efforts to drive research and innovation in systems engineering, tackling complex issues and 5)(提高社会经济

系统的效率和可持续发展能力) _____
_____
_____.

Text B
Navigating Complex Systems

Research Background

Systems engineering is an interdisciplinary field that focuses on the design, integration, and management of complex systems throughout their life cycles. It is applied in various industries such as spacecraft design, computer chip design, robotics, software integration, and bridge building. The use of systems engineering techniques, including modeling and simulation, requirements analysis, and scheduling, helps manage the complexity of these projects. Systems engineering addresses challenges such as requirements engineering, reliability, logistics, coordination of teams, testing and evaluation, and maintainability. It combines technical and human-centered disciplines to ensure all aspects of a project or system are considered and integrated. By applying systems thinking principles, systems engineering aims to create engineered systems that function synergistically to achieve a useful purpose.

Introduction Questions

· How is systems engineering applied to the design, integration, and management of complex projects?

· What are the challenges faced by systems engineering when dealing with large or complex projects?

· What distinguishes the systems engineering process from a manufacturing process?

A. Systems engineering techniques are used in complex projects: spacecraft design, computer chip design, robotics, software integration, and bridge building. Systems engineering uses a host of tools that include modeling and simulation, requirements analysis and scheduling to manage complexity.

B. Systems engineering is an interdisciplinary field of engineering and engineering management that focuses on how to design, integrate, and manage complex systems over their life cycles. At its core, systems engineering utilizes systems thinking principles to organize this body of knowledge. The individual outcome of such efforts, an engineered system, can be defined as a combination of components that work in synergy to collectively perform a useful function.

C. Issues such as requirements engineering, reliability, logistics, coordination of different teams, testing and evaluation, maintainability and many other disciplines necessary for successful system design, development, implementation, and ultimate decommission become more difficult when dealing with large or complex projects. Systems engineering deals with work-processes, optimization methods, and risk management tools in such projects. It overlaps technical and human-centered disciplines such as industrial engineering, production systems engineering, process systems engineering, mechanical engineering, manufacturing engineering, production engineering, control engineering, software engineering, electrical engineering, cybernetics, aerospace engineering, organizational studies, civil engineering and project management.

D. Systems engineering ensures that all likely aspects of a project or system are considered and integrated into a whole.

E. The systems engineering process is a discovery process that is quite unlike a manufacturing process. A manufacturing process is focused on repetitive activities that achieve high quality outputs with minimum cost and time. The systems engineering process must begin by discovering the real problems that need to be resolved, and identify the most probable or highest impact failures that can occur—systems engineering involves finding solutions to these problems.

F. The term systems engineering can be traced back to Bell Telephone Laboratories in the 1940s. The need to identify and manipulate the properties of a system as a whole, which in complex engineering projects may greatly differ from the sum of the parts' properties, motivated various industries, especially those developing systems for the U.S. military, to apply the discipline.

G. When it was no longer possible to rely on design evolution to improve upon a system and the existing tools were not sufficient to meet growing demands, new methods began to be developed that addressed the complexity directly. The continuing evolution of systems engineering comprises the development and identification of new methods and modeling techniques. These methods aid in a better comprehension of the design and developmental control of engineering systems as they grow more complex. Popular tools that are often used in the systems engineering context were developed during these times, including USL, UML, QFD, and IDEF.

H. In 1990, a professional society for systems engineering, the National Council on Systems Engineering (NCOSE), was founded by representatives from a number of U. S. corporations and organizations. NCOSE was created to address the need for improvements in systems engineering practices and education. As a result of growing involvement from systems engineers outside of the U. S., the name of the organization was changed to the International Council on Systems Engineering (INCOSE) in 1995. Schools in several countries offer graduate programs in systems engineering, and continuing education options are also available for practicing engineers.

I. Simon Ramo considered by some to be a founder of modern systems engineering defined the discipline as: "... a branch of engineering which concentrates on the design and application of the whole as distinct from the parts, looking at a problem in its entirety, taking account of all the facets and all the variables and linking the social to the technological."—Conquering Complexity, 2004.

J. "System engineering is a robust approach to the design, creation, and operation of systems. In simple terms, the approach consists of identification and quantification of system goals, creation of alternative system design concepts, performance of design trades, selection and implementation of the best design, verification that the design is properly built and integrated, and post-implementation assessment of how well the system meets (or met) the goals."— NASA Systems Engineering Handbook, 1995.

K. "The Art and Science of creating effective systems, using whole system, whole life principles" OR "The Art and Science of creating optimal solution systems to complex issues and problems"— Derek Hitchins, Prof. of Systems Engineering, former president of INCOSE (UK), 2007.

L. "The systems engineering method recognizes each system is an integrated whole even though composed of diverse, specialized structures and sub-functions. It further recognizes that any system has a number of objectives and that the balance between them may differ widely from system to system. The methods seek to optimize the overall system functions according to the weighted objectives and to achieve maximum compatibility of its parts. " — Systems Engineering Tools by Harold Chestnut, 1965.

M. Systems engineering signifies only an approach and, more recently, a discipline in engineering. The aim of education in systems engineering is to formalize various approaches simply and in doing so, identify new methods and research opportunities similar to that which occurs in other fields of engineering. As an approach, systems engineering is holistic and interdisciplinary in flavour.

Vocabulary List

overlap
to have areas, aspects, or qualities in common or shared between two or more entities　重叠;交叉

compatibility
the ability of two or more systems or components to work together without conflicts　兼容性;适应性

decommission
the act of taking a system, equipment, or facility out of service or operation　停用;废除

quantification
the process of assigning a numerical value or quantity to something　量化;确定……的数量

Language Enhancement

Ⅰ. Choose the most suitable expression in the box to complete each sentence. Change the form where necessary.

compatible	formalize	signify	specialize	manipulate
optimize	integrate	utilize	overlap	verify
synergy	decommission	implement	comprise	coordinate

1. The new software system _____ various data sources to provide a comprehensive solution for data analysis in the field of technology.

2. To improve efficiency, we _____ artificial intelligence algorithms to automate repetitive tasks in our technological processes.

3. The collaboration between engineers and designers created a _____ that resulted in a groundbreaking technological innovation.

4. The project manager _____ the efforts of different teams to ensure smooth execution of the technology project.

5. Our company is ready to _____ the latest cybersecurity measures to protect sensitive information from potential threats.

6. After years of service, the outdated hardware will be _____ to make room for advanced technology in the data center.

7. We continuously _____ our website's performance by analyzing user behavior and implementing speed-enhancing techniques.

8. The functions of the two software applications _____, causing conflicts and compatibility issues in the system.

9. The skilled programmer can _____ the code to achieve the desired functionality in the software application.

10. The new smartphone _____ advanced features, including a high-resolution display, powerful processor, and cutting-edge camera technology.

11. Before deploying the software update, thorough testing is required to _____ its compatibility with different operating systems.

12. In the field of artificial intelligence, many researchers _____ in natural language processing or computer vision.

13. The use of standard protocols in networking systems _____ interoperability between different devices and technologies.

14. The new software update is _____ with multiple operating systems, allowing users to seamlessly access it on various devices.

15. The team of engineers worked together to design and _____ the specifications of the new technological device.

Ⅱ. **Choose the most suitable expression in the box to complete each sentence. Change the form where necessary.**

integrate into	trace back to	a host of	distinct from	compose of
concentrate on	differ from	focus on	take account of	

1. _____ cutting-edge technologies, such as artificial intelligence, blockchain, and quantum computing, are revolutionizing multiple industries simultaneously.

2. The research project _____ developing advanced algorithms and computational models to enhance the accuracy and efficiency of weather prediction systems.

3. The new software aims to seamlessly _____ existing enterprise resource planning systems, enabling smooth data synchronization and process automation.

4. By _____ the origins of programming languages, one can understand the evolution of software development and the foundations of modern coding practices.

5. The characteristics of renewable energy sources _____ those of fossil fuels, as they offer sustainability, lower emissions, and reduced dependency on finite resources.

6. In order to achieve breakthroughs in space exploration, scientists and engineers _____ developing propulsion systems, spacecraft materials, and advanced navigation technologies.

7. The approach to user interface design in mobile applications is _____ that in web development, considering factors such as touch gestures, screen size, and device capabilities.

8. When designing a smart city infrastructure, urban planners must _____ various factors, including population density, transportation systems, energy consumption, and environmental sustainability.

9. The composition of a multidisciplinary research team, _____ experts in biology, computer science, and chemistry, ensures a holistic approach to tackling complex biomedical challenges.

Ⅲ. Complete the summary using the suitable words in the Text B.

Systems engineering is an interdisciplinary field of engineering that is applied to address complex projects. It 1) _____ various 2) _____ such as spacecraft design, computer chip design, robotics, software integration, and bridge construction. Systems engineering 3) _____ tools like modeling, 4) _____, requirements analysis, and scheduling to manage complexity. At the core of this field is the application of systems thinking principles to organize knowledge. The outcome of systems engineering is an engineered system, where the components work together synergistically to 5) _____ a useful function.

In large or complex projects, systems engineering faces numerous challenges in disciplines such as requirements engineering, reliability, logistics coordination, testing and evaluation, and maintainability. It involves work processes, 6) _____ methods, and risk management tools to overcome these 7) _____. Systems engineering bridges technical and human-centered 8) _____ including industrial engineering, manufacturing engineering, software engineering, and electrical engineering. It ensures that all aspects of a project or system are considered and 9)＿＿＿＿ into a cohesive whole.

Systems engineering is a discovery process 10)＿＿＿＿ from manufacturing processes. It involves identifying and 11)＿＿＿＿ real problems, as well as finding solutions to potential failures. The development of systems engineering 12)＿＿＿＿ from the need to understand and manipulate the properties of a system as a whole, particularly in the field of U. S. military system development. It has continued to 13)＿＿＿＿, encompassing the development and 14)＿＿＿＿ of new methods and modeling techniques to better 15)＿＿＿＿ and control increasingly complex engineering systems.

Academic Expression

Pair work: Discuss with your partner and compare the two possible paraphrases of each sentence and decide which one expresses the original meaning more academically.

1. The systems engineering process must begin by discovering the real problems that need to be resolved, and identify the most probable or highest impact failures that can occur.

    a. The first step in the systems engineering process is figuring out the actual problems that need fixing and pinpointing the failures that are most likely to happen or have the biggest impact.

    b. The systems engineering process initiates by identifying the underlying problems that necessitate resolution and determining the most likely or significant potential failures.

2. The continuing evolution of systems engineering comprises the development and identification of new methods and modeling techniques.

    a. The progressive advancement of systems engineering encompasses the formulation and recognition of novel methodologies and modeling techniques.

    b. Systems engineering keeps evolving by coming up with and discovering new methods and ways to model things.

3. As a result of growing involvement from systems engineers outside of the U. S., the name of the organization was changed to the International Council on Systems Engineering (INCOSE) in 1995.

    a. Because more systems engineers from other countries got involved, they decided to change the organization's name to the International Council on Systems Engineering (INCOSE) in 1995.

b. Due to increased participation of international systems engineers, the organization's name was modified to the International Council on Systems Engineering (INCOSE) in 1995.

4. Systems engineering signifies only an approach and, more recently, a discipline in engineering.

a. Systems engineering basically means a way of doing things and, more recently, it's become a recognized field of engineering.

b. Systems engineering denotes primarily an approach and, more recently, an engineering discipline.

5. The aim of education in systems engineering is to formalize various approaches simply and in doing so, identify new methods and research opportunities similar to that which occurs in other fields of engineering.

a. The objective of systems engineering education is to systematically formalize diverse approaches and, in the process, identify novel methods and research prospects analogous to those found in other engineering disciplines.

b. The goal of studying systems engineering is to make different approaches more organized and, by doing that, discover new methods and research opportunities, just like they do in other engineering fields.

6. The individual outcome of such efforts, an engineered system, can be defined as a combination of components that work in synergy to collectively perform a useful function.

a. The resultant product of these endeavors, namely an engineered system, can be described as an amalgamation of interdependent components that synergistically collaborate to collectively execute a beneficial function.

b. The end result of all this hard work is an engineered system, which is basically a combination of different parts working together to do something useful.

7. The methods seek to optimize the overall system functions according to the weighted objectives and to achieve maximum compatibility of its parts.

a. The methods are designed to make the whole system work as efficiently as possible, taking into account the most important goals and ensuring that all the different parts fit together seamlessly.

b. The methodologies aim to optimize the holistic system functionalities in accordance with prioritized objectives and attain maximum compatibility among its constituent parts.

8. Systems engineering techniques are used in complex projects: spacecraft design, computer chip design, robotics, software integration, and bridge building.

  a. Systems engineering methodologies are employed in intricate endeavors encompassing spacecraft design, computer chip design, robotics, software integration, and bridge construction.

  b. Systems engineering methods are applied in all sorts of complicated projects like designing spaceships, computer chips, robots, putting software together, and building bridges.

Understanding the Text

Ⅰ. **Text B has twelve paragraphs, A-L. Which paragraph contains the following information? You may use any letter more than once.**

1. Simon Ramo, a key figure in systems engineering, defined it as an engineering branch that considers problems as a whole, encompassing social and technological aspects.

2. Systems engineering utilizes a range of tools such as modeling and simulation, requirements analysis, and scheduling to manage complexity.

3. The systems engineering process differs from a manufacturing process as it focuses on discovering and solving real problems rather than repetitive activities.

4. Systems engineering deals with work-processes, optimization methods, and risk management tools to address these challenges.

5. The International Council on Systems Engineering (INCOSE) was established in 1990 to improve systems engineering practices and education.

6. The term "systems engineering" originated in the 1940s, driven by the need to understand and manipulate the properties of complex systems.

7. Large or complex projects present challenges in areas like requirements engineering, reliability, logistics coordination, testing and evaluation, and maintainability.

8. The evolution of systems engineering involves the development of new methods and modeling techniques to handle increasing complexity.

9. Systems engineering techniques are applied in complex projects, including spacecraft design, computer chip design, robotics, software integration, and bridge building.

10. Systems engineering is a robust approach that involves identifying system goals, creating design concepts, performing design trades, implementing the best design, and assessing its performance.

II. Do the following statements agree with the information given in Text B? Write your judgement.

TRUE   if the statement agrees with the information
FALSE   if the statement contradicts the information
NOT GIVEN   if there is no information on this

1. Systems engineering somewhat ensures that most aspects of a project or system are considered and integrated into a whole.
2. The final result of these efforts is an engineered system, which refers to a combination of components working together synergistically to accomplish a practical purpose.
3. The ongoing development and identification of new methods and modeling techniques are part of the continuous evolution of systems engineering.
4. Graduate programs in systems engineering are offered by universities in multiple countries, and practicing engineers can also pursue continuing education opportunities.
5. Systems engineering utilizes advanced computational methods to organize this body of knowledge.
6. Systems engineering techniques are only used in computer chip design and robotics, not in spacecraft design, software integration, or bridge building.
7. Systems engineering involves finding innovative solutions to these problems.
8. Systems engineering only deals with work-processes in such projects, excluding optimization methods and risk management tools.
9. The systems engineering process is somewhat different from a manufacturing process.
10. Representatives from various U. S. corporations and organizations established the

Unit Six　Systems Engineering

National Council on Systems Engineering (NCOSE), a professional society for systems engineering, in 1990.

III. Group work: Please work in groups and choose/use no more than three words to complete the structural table.

Section	Paragraph	Contents
Introduction and Definition	E	Introduction to the term "systems engineering" and its origin.
	B	Definition and focus of systems engineering as 1) _____.
Applications and Techniques	A	Examples of 2) _____ where systems engineering techniques are used.
	C	3) _____ and disciplines in large or complex projects and the role of systems engineering.
	D	Comparison of the systems engineering process with a manufacturing process.
Evolution and Development	F	Evolution of systems engineering and development of new methods and 4) _____.
Professional Society and Education	G	Establishment of the National Council on Systems Engineering (NCOSE) and its 5) _____ the International Council on Systems Engineering (INCOSE).
	L	Overview of systems engineering as an 6) _____ in engineering and its aim in education.
Perspectives and Definitions	H	Definition of systems engineering by 7) _____.
	I	Description of the systems engineering approach and its 8) _____.
	J	Quotes defining systems engineering as the art and science of creating effective and 9) _____.
	K	Explanation of the systems engineering method and 10) _____.

Rise of a Great Power

Translate the following Chinese text into English.

In the development of Chinese aircraft carriers, systems engineering plays a crucial role. As a complex military project, 1)(航母的设计、建造和运营需要协调和整合众多的系统和子系统,而系统工程正是在这个过程中发挥着关键的作用。)_____

_____.

An aircraft carrier is a vast system 2)(包括动力系统、通信系统、导航系统、武器系统等多个子系统。)_____

_____.

These subsystems need to work closely together to ensure that the aircraft carrier can accomplish complex tasks and combat operations. 3)(系统工程师通过整合和协调这些子系统,使它们能够有效地相互配合,)_____

_____ thereby improving the overall system's performance and reliability. Moreover, any malfunction or lack of coordination in any system of an aircraft carrier can significantly impact its overall performance. Systems engineers, through detailed system analysis and simulation experiments, 4)(能够提前发现和解决潜在的问题,减少项目失败的可能性)_____

_____.

Systems engineering also focuses on 5)(设计易维护操作系统,降低成本,提高可用性)_____

_____.

In the ongoing development of China's aircraft carrier program, systems engineering continues to provide technical support and drive progress.

6-1 Unit Six-Answer and Translation

Unit Seven Robotics

Warm-up

Implementing the strategy of building a strong manufacturing country is in line with the fundamental requirements of the "two centenary goals" and the "Chinese Dream". To achieve these goals, there must be a solid economic foundation and a strong manufacturing industry to support them, which urgently requires China's manufacturing industry to transform and upgrade, better meet the needs of economic and social development and national defense construction.

Thought-provoking Questions

· What are the key challenges and obstacles that China's manufacturing industry faces in its transformation and upgrade process?

· How can the transformation and upgrade of China's manufacturing industry contribute to the realization of the "two centenary goals" and the "Chinese Dream"?

Text A
The Transformative Field of Robotics

Research Background

Robotics is a fascinating and rapidly advancing field that encompasses the design, construction, operation, and utilization of robots. The word "robotics" has a rich history, originating from different sources and gaining prominence through various works of literature and science fiction. With the integration of multiple disciplines, robotics has found practical applications in industries, hazardous environments, and areas where human presence is impractical or unsafe. The development of autonomous robots, inspired by nature and capable of replicating human actions, is a key focus in this field.

Introduction Questions

· What is the etymology and historical background of the word "robotics"?
· What is the historical significance of the term "robotics" in science fiction literature?
· What are the key domains and applications of robotics?

A. The word robotics was derived from the word robot, which was introduced to the public by Czech writer Karel Čapek in his play R. U. R. (Rossum's Universal Robots), which was published in 1920. The word robot comes from the Slavic word robota, which means work/job. The play begins in a factory that makes artificial people called robots, creatures who can be mistaken for humans-very similar to the modern ideas of androids. Karel Čapek himself did not coin the word. He wrote a short letter in reference to an etymology in the Oxford English Dictionary in which he named his brother Josef Čapek as its actual originator.

B. According to the Oxford English Dictionary, the word robotics was first used in print by Isaac Asimov, in his science fiction short story "Liar!", published in May 1941 in Astounding

Science Fiction. Asimov was unaware that he was coining the term; since the science and technology of electrical devices is electronics, he assumed robotics already referred to the science and technology of robots. In some of Asimov's other works, he states that the first use of the word robotics was in his short story Runaround (Astounding Science Fiction, March 1942), where he introduced his concept of The Three Laws of Robotics. However, the original publication of "Liar!" predates that of "Runaround" by ten months, so the former is generally cited as the word's origin.

C. In 1948, Norbert Wiener formulated the principles of cybernetics, the basis of practical robotics. Fully autonomous robots only appeared in the second half of the 20th century. The first digitally operated and programmable robot, the Unimate, was installed in 1961 to lift hot pieces of metal from a die casting machine and stack them. Commercial and industrial robots are widespread today and used to perform jobs more cheaply, more accurately, and more reliably than humans. They are also employed in some jobs which are too dirty, dangerous, or dull to be suitable for humans. Robots are widely used in manufacturing, assembly, packing and packaging, mining, transport, earth and space exploration, surgery, weaponry, laboratory research, safety, and the mass production of consumer and industrial goods.

D. Robotics, an interdisciplinary branch of computer science and engineering, encompasses the design, construction, operation, and utilization of robots. The primary goal of robotics is to create machines that can assist and collaborate with humans in various tasks. This field integrates multiple domains such as mechanical engineering, electrical engineering, information engineering, mechatronics engineering, electronics, biomedical engineering, computer engineering, control systems engineering, software engineering, and mathematics. By developing machines capable of replicating human actions, robotics has found applications in hazardous environments, manufacturing processes, and areas where human presence is impractical or unsafe.

E. Robotics develops machines that can substitute for humans and replicate human actions. Robots can be used in many situations for many purposes, but today many are used in dangerous environments (including inspection of radioactive materials, bomb detection and deactivation), manufacturing processes, or where humans cannot survive (e. g. , in space, underwater, in high heat, and clean up and containment of hazardous materials and radiation). Robots can take any form, but some are made to resemble humans in appearance. This is claimed to help in the acceptance of robots in certain replicative behaviors which are usually performed by people. Such robots attempt to replicate walking, lifting, speech, cognition, or any other tasks mainly performed by a human. Many of today's robots are inspired by nature, contributing to the field of bio-inspired robotics.

F. Certain robots require user input to operate, while other robots function autonomously. The concept of creating robots that can operate autonomously dates back to classical times, but research into the functionality and potential uses of robots did not grow substantially until the 20th century. Throughout history, it has been frequently assumed by various scholars, inventors, engineers, and technicians that robots will one day be able to mimic human behavior and manage tasks in a human-like fashion. Today, robotics is a rapidly growing field, as technological advances continue; researching, designing, and building new robots serve various practical purposes, whether domestically, commercially, or militarily. Many robots are built to do jobs that are hazardous to people, such as defusing bombs, finding survivors in unstable ruins, and exploring mines and shipwrecks. Robotics is also used in STEM (science, technology, engineering, and mathematics) as a teaching aid.

G. With its integration of mechanical engineering, electrical engineering, information engineering, mechatronics engineering, and other disciplines, robotics plays a crucial role in shaping the future of automation and technological advancements. As the field continues to grow, researchers and engineers strive to develop robots that can mimic human behavior, perform complex tasks, and enhance human life in diverse domains.

Vocabulary List

robotics
the branch of technology that deals with robots   机器人技术

androids
humanoid robots   人形机器人

etymology
the study of the origin and history of words   词源学

deactivation
the process of making something inactive   停用,失效

containment
the action of keeping something within limits   防护,遏制

replicative

relating to replication or reproduction 复制的,复制品的

cognition

the mental process of acquiring knowledge 认知

mimic

to imitate or copy the actions or speech of someone 模仿

defuse

to reduce the danger or tension in a situation 缓解,平息

Language Enhancement

Ⅰ. Choose the most suitable word in the box below to complete each sentence. Change the form where necessary.

resemble	mimic	enhance	coin	programmable
cognition	defuse	predate	utilize	substitute
inspire	formulate	replicate	deactivation	containment

1. The term "artificial intelligence" was _____ by John McCarthy in 1956.
2. Ancient civilizations _____ modern technology by thousands of years.
3. Scientists are working hard to _____ a new theory of quantum mechanics.
4. The _____ robot can be customized to perform various tasks.
5. The _____ of renewable energy sources is crucial for a sustainable future.
6. The scientists were able to _____ the experiment's results multiple times.
7. In the absence of the required ingredient, you can _____ it with a similar alternative.
8. The _____ of the security system allowed unauthorized access to the data.
9. Strict _____ measures were implemented to prevent the spread of the infectious disease.
10. The newly developed device closely _____ its predecessor in terms of design and functionality.
11. _____ abilities play a crucial role in problem-solving and decision-making.

12. The innovative ideas presented by the speaker _____ the audience to pursue technological advancements.

13. The robot was designed to _____ human gestures and facial expressions.

14. The bomb squad worked diligently to _____ the explosive device safely.

15. The new software update is expected to _____ the performance and efficiency of the system.

Ⅱ. **Choose the most suitable expression in the box to complete each sentence. Change the form where necessary.**

refer to	be mistaken for	strive to	be similar to	date back to
attempt to	in reference to	derive from	collaborate with	play a crucial role in

1. The concept of artificial intelligence _____ the early work of Alan Turing.

2. The prototype robot was so lifelike that it was often _____ a human.

3. The interface of the new operating system is designed to _____ that of its predecessor.

4. _____ the previous study, the researchers conducted additional experiments to validate the findings.

5. Data analysis _____ extracting meaningful insights from large datasets.

6. When discussing advanced technology, "nanotechnology" often _____ the manipulation of materials at the atomic level.

7. The company decided to _____ a renowned research institution to develop innovative solutions.

8. The engineers _____ optimize the efficiency of the solar panels through various design modifications.

9. The origins of computer programming _____ the 19th century with the invention of the Analytical Engine by Charles Babbage.

10. The research team _____ push the boundaries of knowledge in the field of quantum computing.

Ⅲ. **Skim the text and complete the summary using the list of words in the box. Change the form where necessary.**

hazardous	replicate	collaborate	association	utilization
strive	touch	formulation	exploration	combine
mimic	autonomous	interdisciplinary	domains	enhance

This passage provides an overview of robotics and its historical development. It begins by explaining the origin of the word "robotics" and its connection to the term "robot", which was introduced by Karel Čapek in his play "R. U. R." The passage then discusses the early usage of the word "robotics" by Isaac Asimov and its 1)_____ with the science and technology of robots. It highlights the 2)_____ of cybernetics by Norbert Wiener and the emergence of fully 3)_____ robots in the latter half of the 20th century.

The passage emphasizes the widespread 4)_____ of robots in various industries, highlighting their ability to perform tasks more efficiently and safely than humans. It mentions their applications in manufacturing, mining, 5)_____, surgery, weaponry, and other fields. Robotics is described as an 6)_____ branch that 7)_____ multiple 8)_____ such as mechanical engineering, electrical engineering, computer science, and mathematics.

Furthermore, the passage explains that robotics aims to create machines that can 9)_____ with humans in different tasks. It mentions the importance of robots in 10)_____ environments and their ability to 11)_____ human actions. The role of bio-inspired robotics is also 12)_____ upon. The passage concludes by highlighting the growing significance of robotics in automation and technological advancements, as researchers and engineers 13)_____ to develop robots that can 14)_____ human behavior and 15)_____ various aspects of human life.

Academic Expression

Pair work: Discuss with your partner and compare the two possible paraphrases of each sentence and decide which one expresses the original meaning more academically.

1. The word "robotics" was derived from the word "robot".
    a. The term "robotics" was etymologically derived from the term "robot".
    b. The word "robotics" came from the word "robot".

2. Karel Čapek wrote a short letter in reference to an etymology in the Oxford English Dictionary.
    a. Karel Čapek wrote a short letter referring to an explanation in the Oxford English Dictionary.
    b. Karel Čapek authored a brief correspondence alluding to an etymology mentioned in the Oxford English Dictionary.

3. Robots are widely used in manufacturing, assembly, packing and packaging, mining, transport, earth and space exploration, surgery, weaponry, laboratory research, safety, and the mass production of consumer and industrial goods.

  a. Robots find extensive applications in diverse domains such as manufacturing, assembly, packaging, mining, transportation, space and terrestrial exploration, surgical procedures, weaponry, laboratory research, safety measures, and the large-scale production of consumer and industrial commodities.

  b. Robots are used in a wide range of industries, including manufacturing, assembly, packaging, mining, transportation, space exploration, surgery, weapons development, laboratory research, safety measures, and the mass production of consumer and industrial goods.

4. Robotics encompasses the design, construction, operation, and utilization of robots.
  a. Robotics involves designing, building, operating, and using robots.
  b. Robotics entails the comprehensive aspects of robot design, fabrication, functionality, and application.

5. Robots can be used in many situations for many purposes.
  a. Robots possess versatile utility across various contexts and objectives.
  b. Robots can be used in a lot of different situations for many different purposes.

6. The concept of creating robots that can operate autonomously dates back to classical times.
  a. The notion of developing self-operating robots traces its origins to antiquity.
  b. The idea of making robots that can work on their own goes way back to ancient times.

7. Researching, designing, and building new robots serve various practical purposes.
  a. Researching, designing, and building new robots have many practical purposes.
  b. The pursuit of research, design, and construction of novel robots serves a multitude of pragmatic objectives.

8. Robotics plays a crucial role in shaping the future of automation and technological advancements.
  a. Robotics is extremely important in shaping the future of automation and technological advancements.

b. Robotics assumes a pivotal role in shaping the trajectory of automation and technological progress.

Understanding the Text

I. Pair work: Work with your partner and take turns asking and answering the following questions according to the information contained in the text.

1. Who coined the word "robotics"?

2. What is the origin of the word "robot"?

3. Which play introduced the word "robot" to the public?

4. What is the primary goal of robotics?

5. What are some areas where robots are widely used today?

6. Which interdisciplinary fields are integrated into robotics?

7. What are some tasks that robots are designed to perform in hazardous environments?

8. What role does robotics play in shaping the future of automation and technological advancements?

II. Group work: Please work in groups and use no more than three words to complete the structural table according to the text.

Section	Paragraph	Content
Introduction	A	The word "robotics" comes from "robot" introduced by Karel Čapek in his play R. U. R. in 1920. It refers to 1)_____ called robots. Čapek credited his brother Josef Čapek as its originator.

Continued Table

Section	Paragraph	Content
Main Body	B	According to the Oxford English Dictionary, Isaac Asimov first used "robotics" in his story "Liar!" published in 1941. He assumed it referred to 2)_____, later mentioning "Runaround" as the first use. However, "Liar!" is commonly considered the 3)_____.
	C	Norbert Wiener 4)_____ cybernetics principles in 1948, forming the basis of practical robotics. Autonomous robots emerged in the second half of the 20th century. The Unimate, installed in 1961, was the first 5)_____ used in manufacturing and industrial tasks.
	D	Robotics is an 6)_____ of computer science and engineering. Its goal is to create robots that assist humans in various tasks. It integrates mechanical and electrical engineering, software engineering, and mathematics.
	E	Robotics develops machines that replace humans in dangerous, manufacturing, and 7)_____. Some robots resemble humans to 8)_____ human actions. They are used in tasks such as radioactive material inspection, bomb detection, and hazardous material cleanup.
	F	Robots can operate with user input or autonomously. Research on autonomous robots increased in the 20th century, aiming to 9)_____. Robotics is a growing field with applications in areas like bomb demolition, search and rescue, and education.
Conclusion	G	Robotics, integrating various disciplines, plays a vital role in automation and technological advancements. Researchers strive to create robots that mimic human behavior and 10)_____ in different domains.

Rise of a Great Power

Translate the following Chinese part into English and make it a complete English text.

Driven by economic development and technological innovation, 1)(中国的机器人产业正

迅速崛起,并成为全球机器人研究和应用的重要力量)_____
_____
_____. 2)(中国的机器人应用领域广泛),
_____,
including manufacturing, healthcare, agriculture, and logistics. Robots play a crucial role in enhancing production efficiency, labor substitution, and 3)(解决人口老龄化)_____
_____. For instance, in the realm of industrial manufacturing, robot deployment has facilitated automated production, leading to improved product quality and productivity. In the field of healthcare, robots assist in surgeries and provide rehabilitation treatments, 4)(改善了医疗水平和患者体验)_____
_____.

Looking ahead, Through continuous dedication and collaboration on a global scale, 5)(中国的机器人产业将继续迈向新的高度,为构建智能化、高效率的社会做出重要贡献)_____
_____
_____.

Text B
Autonomous Modular Robot System
Tasks, Adaptability, and Future Improvements

Research Background

This paper presents a modular robot system that autonomously completes high-level tasks by reactively reconfiguring in response to its perceived environment and task requirements. Hardware demonstrations reveal opportunities for improvement. MSRRs (Modular self-reconfigurable robot systems) are mechanically distributed, allowing for distributed planning, sensing, and control. However, our system differs by being distributed at the hardware level and centralized at the high level, leveraging advantages of both paradigms. SMORES-EP demonstrates various interactions with environments and objects through reconfiguration. Centralized sensing and control enable rapid transformations between configurations. Challenges include robustness and cascading failures, but future steps involve incorporating more feedback and exploring distributed repair strategies.

Introduction Questions

· What is the main focus of this paper on modular robot systems?

· How does the system in this research differ from past modular self-reconfigurable robot systems?

· What are the challenges observed in the hardware demonstrations and potential areas for improvement in future systems?

A. This paper presents a modular robot system that autonomously completed high-level tasks by reactively reconfiguring in response to its perceived environment and task requirements. Putting the entire system to the test in hardware demonstrations revealed several opportunities for future improvement. MSRRs(Modular self-reconfigurable robot (MSRR) systems) are by their nature mechanically distributed and, as a result, lend themselves naturally to distributed planning, sensing, and control. Most past systems have used entirely distributed frameworks. Our system was designed differently. It is distributed at the low level (hardware) but centralized at the high level (planning and perception), leveraging the advantages of both design paradigms.

B. The three scenarios in the demonstrations showcase a range of different ways SMORES-EP(Self-Assembling Modular Robot for Extreme Environments with Payloads) can interact with environments and objects: moving over flat ground, fitting into tight spaces, reaching up high, climbing over rough terrain, and manipulating objects. This broad range of functionality is only accessible to SMORES-EP by reconfiguring between different morphologies.

C. The high-level planner, environment characterization tools, and library worked together to allow tasks to be represented in a flexible and reactive manner. For example, at the high level, demonstrations II and III were the same task: deliver an object at a goal location. However, after characterizing the different environments (high in II, stairs in III), the system automatically determined that different configurations and behaviors were required to complete each task: the Proboscis to reach up high, and the Snake to climb the stairs. Similarly, in demonstration I, there was no high-level distinction between the green and pink objects—the robot was simply asked to retrieve all objects it found. The sensed environment once again dictated the choice of behavior: the simple problem (object in the open) was solved in a simple way (with the Car configuration), and the more difficult problem (object in tunnel) was solved in a more sophisticated way (by reconfiguring into the Proboscis). This level of sophistication in control and decision-making goes beyond the capabilities demonstrated by past systems with distributed architectures.

D. Centralized sensing and control during reconfiguration, provided by AprilTags and a centralized path planner, allowed our implementation to transform between configurations more rapidly than previous distributed systems. Each reconfiguration action (a module disconnecting, moving, and reattaching) takes about 1 min. In contrast, past systems that used distributed sensing and control required 5 to 15 min for single-reconfiguration actions, which would prohibit their use in the complex tasks and environments that our system demonstrated.

E. Through the hardware demonstrations performed with our system, we observed several challenges and opportunities for future improvement. All SMORES-EP body modules are identical and therefore interchangeable for the purposes of reconfiguration. However, the sensor module has a substantially different shape than a SMORES-EP body module, which introduces heterogeneity in a way that complicates motion planning and reconfiguration planning. Configurations and behaviors must be designed to provide the sensor module with an adequate view and to support its weight and elongated shape. Centralizing sensing also limits reconfiguration: modules can only drive independently in the vicinity of the sensor module, preventing the robot from operating as multiple disparate clusters.

F. Our high-level planner assumes that all underlying components are reliable and robust, so failure of a low-level component can cause the high-level planner to behave unexpectedly and result in failure of the entire task. A study shows the causes of failure for 24 attempts of demonstration (placing the stamp on the package). Nearly all failures were due to an error in one of the low-level components that the system relies on, with 42% of failure due to hardware errors and 38% due to failures in low-level software (object recognition, navigation, and environment characterization). This kind of cascading failure is a weakness of centralized, hierarchical systems: Distributed systems are often designed so that failure of a single unit can be compensated for by other units and does not result in global failure.

G. This lack of robustness presents a challenge, but steps can be taken to address it. Open-loop behaviors (such as stair climbing and reaching up to place the stamp) were vulnerable to small hardware errors and less robust against variations in the environment. For example, if the height of stairs in the actual environment is higher than the property value of the library entry, then the stair-climbing behavior is likely to fail. Closing the loop using sensing made exploration and reconfiguration significantly less vulnerable to error. Future systems could be made more robust by introducing more feedback from low-level components to high-level decision-making processes and by incorporating existing high-level failure-recovery frameworks (24). Distributed repair strategies could also be explored, to replace malfunctioning modules with nearby working ones on the fly.

H. To implement our perception characterization component, we assumed a limited set of environment types and implemented a simple characterization function to distinguish between them. This function does not generalize very well to completely unstructured environments and also is not very scalable. Thus, to expand the system to work well for more realistic environments and to distinguish between a large number of environment types, a more general characterization function should be implemented.

Vocabulary List

modular
consisting of separate modules or units that can be combined or rearranged 模块化的,由可组合或重新排列的独立模块或单元组成

reconfiguring
rearranging or reorganizing a system or structure 重新配置,重新组织系统或结构

paradigms
models or patterns that serve as examples or standards 范例,模式

morphology
the study of the form and structure of organisms or objects 形态学,结构形态学

configuration
the arrangement or setting of elements within a system or structure 配置,结构

proboscis
a long flexible snout or trunk-like structure 长鼻,喙

heterogeneity

the quality or state of being diverse or composed of different elements   多样性,异质性

elongated

extended or lengthened in shape   延长的,伸长的

vicinity

the area or region near or surrounding a particular place   附近,周围地区

disparate

distinct or different from each other   迥然不同的,不相同的

clusters

groups or collections of similar things or individuals   集群,群集

cascading

progressing or spreading in a sequence or chain reaction   级联的,级联递进的

malfunctioning

not working or operating properly   故障,发生故障

scalable

capable of being easily expanded or adapted to a larger size or scope   可扩展的,可伸缩的

Language Enhancement

Ⅰ. **Choose the most suitable word in the box below to complete each sentence. Change the form where necessary.**

disparate	distinguish	centralize	interchangeable	implement
incorporate	reconfigure	transform	dictate	sophisticate

1. The modular robot system can _____ itself autonomously in response to changes in its environment.

2. The _____ control unit allows the system to efficiently manage and coordinate the actions of multiple robots.

157

3. The requirements of the task _____ the specific configuration and behavior of the robot.

4. The _____ algorithm implemented in the system enables complex decision-making and problem-solving capabilities.

5. The successful _____ of the new technology transformed the industry, improving efficiency and productivity.

6. The _____ modules can be easily swapped and reconfigured to adapt to different tasks and scenarios.

7. The system _____ advanced sensors and AI algorithms to enhance perception and decision-making abilities.

8. The advanced facial recognition technology can accurately _____ between identical twins based on their unique facial features.

9. The _____ data sources were consolidated and centralized, enabling better analysis and insights.

10. The team worked diligently to _____ user feedback into the design of the software, resulting in a more user-friendly interface.

Ⅱ. **Choose the suitable phrases in the box below to complete each sentence. Change the form where necessary.**

rely on	in response to	be vulnerable to	fit into
in contrast to	be due to	result in	

1. The experiment showed that the combination of two chemicals _____ a powerful reaction.

2. As technology advances, our personal information becomes more _____ cyber attacks.

3. The delay in project completion _____ a shortage of skilled workers.

4. The success of the mission _____ the precise coordination of multiple teams.

5. _____ customer feedback, the company introduced a new and improved version of their product.

6. _____ traditional methods, the new software streamlines the entire process.

7. The puzzle pieces _____ each other perfectly, creating a complete picture.

Ⅲ. **Skim the text and complete the summary using the list of words in the box.**

flexible	distributed	reliance	requirements	incorporating
efficiency	interacting	enhancing	reconfiguring	sensor
heterogeneity	environments	adaptability	autonomously	robustness

This paper introduces a modular robot system that 1)_____ completes high-level tasks by 2)_____ in response to the environment and task 3)_____. The system combines 4)_____ planning and control, leveraging both centralized and distributed design paradigms. Demonstrations showcase the system's capabilities in 5)_____ with various 6)_____ and objects. The high-level planner, environment characterization tools, and library enable 7)_____ task representation. Centralized sensing and control during reconfiguration improve 8)_____ compared to previous systems. Challenges include the 9)_____ introduced by the 10)_____ module and the 11)_____ on reliable low-level components. Steps to address these issues involve 12)_____ feedback, introducing 13)_____ measures, and exploring 14)_____ repair strategies. Additionally, the perception characterization component needs further development to handle unstructured environments. Future improvements should focus on enhancing the system's scalability and 15)_____ to diverse environments.

Academic Expression

Pair work: Discuss with your partner and compare the two possible paraphrases of each sentence and decide which one expresses the original meaning more academically.

1. The high-level planner, environment characterization tools, and library worked together to allow tasks to be represented in a flexible and reactive manner.

    a. The high-level planner, environment characterization tools, and library joined forces to make tasks adaptable and responsive.

    b. The synergy among the high-level planner, tools for environment characterization, and library facilitated the representation of tasks with flexibility and responsiveness.

2. Through the hardware demonstrations performed with our system, we observed several challenges and opportunities for future improvement.

    a. By conducting hardware demonstrations using our system, we identified numerous challenges and prospects for further enhancements.

    b. While testing our system with actual hardware, we encountered a bunch of challenges and spotted exciting possibilities for making it even better.

3. This lack of robustness presents a challenge, but steps can be taken to address it.

    a. The deficiency in robustness poses a significant challenge; however, mitigating measures can be implemented to tackle this issue.

b. The system's weakness is a bit of a hurdle, but don't worry, we have ideas to overcome it.

4. Nearly all failures were due to an error in one of the low-level components that the system relies on, with 42% of failure due to hardware errors and 38% due to failures in low-level software.

　　a. Almost all the failures happened because of glitches in the system's basic parts, like hardware or software, with hardware errors causing 42% and low-level software screw-ups contributing to 38% of the failures.

　　b. The majority of failures stemmed from errors occurring in the low-level components, upon which the system is dependent, with hardware errors accounting for 42% of the failures and low-level software failures constituting 38%.

5. To expand the system to work well for more realistic environments and to distinguish between a large number of environment types, a more general characterization function should be implemented.

　　a. In order to enhance the system's compatibility with diverse and realistic environments, as well as its ability to differentiate between numerous environmental types, it is recommended to introduce a more comprehensive characterization function.

　　b. To make the system rock in real-life situations and handle various environments like a champ, we gotta upgrade the way it understands different types of surroundings.

Understanding the Text

Ⅰ. **Text B has eight paragraphs, A-H. Which paragraph contains the following information? You may use any letter more than once.**

1. More feedback from low-level components to high-level decision-making processes could make future systems more robust.

2. Failure of a low-level component can cause the high-level planner to behave unexpectedly and result in failure of the entire task.

3. Centralized sensing and control during reconfiguration allow rapid transformation between configurations.

4. SMORES-EP can interact with environments and objects in different ways by reconfiguring between different morphologies.

5. A more general characterization function should be implemented to expand the system's capability in realistic environments.

6. The system automatically determines different configurations and behaviors based on the perceived environment and task requirements.

7. Our system is distributed at the low level (hardware) but centralized at the high level (planning and perception).

8. Distributed repair strategies could be explored to replace malfunctioning modules with nearby working ones.

9. The heterogeneity introduced by the sensor module complicates motion planning and reconfiguration planning.

10. Open-loop behaviors are vulnerable to small hardware errors and variations in the environment.

II. Do the following statements agree with the information given in Text B? Write your judgement.

YES	*if the statement agrees with the information*
NO	*if the statement contradicts the information*
NOT GIVEN	*if there is no information on this*

1. A new modular robot system completes tasks autonomously by reacting to the environment and requirements, combining distributed and centralized design.

2. The reconfiguration between different morphologies does not effectively enhance its capabilities as expected.

3. The complexity of control and decision-making demonstrated by this system is similar to that of past systems with distributed architectures.

4. The high-level planner, environment characterization tools, and library collaborated seamlessly, allowing the system to adaptively learn and optimize task representations.

5. Centralized sensing and control enabled faster reconfiguration, outperforming distributed systems in complex tasks.

6. For reconfiguration, all body modules of SMORES-EP are similar and can be interchanged.

7. In the case of errors in low-level components, the failure rate of hardware is higher than that of low-level software.

8. Behaviors, which rely on open-loop control, were susceptible to minor hardware errors and resilient to environmental variations.

Ⅲ. **Group work**: Please work in groups and complete the structural table with the information from Text B.

Function	Paragraph	Main Content
Introduction	A. Introduction and Background	Introduces the 1)_____ and 2)_____ of the modular robot system, emphasizing the advantages of distributed planning and control.
	B. Demonstrations and Functionality	Showcases the different ways SMORES-EP interacts with environments and objects, highlighting the capability achieved through 3)_____.
Body	C. Coordination and Task Configuration	Explains the 4)_____ and responsiveness of the high-level planner, environment characterization tools, and library in adapting to different task configurations based on the 5)_____.
	D. Advantages of Centralized Sensing and Control	Discusses the role of centralized sensing and control in facilitating faster reconfiguration compared to previous 6)_____ systems.
	E. Challenges and Improvement Opportunities from Hardware Demonstrations	Observes the challenges and improvement opportunities faced during hardware demonstrations, particularly regarding the impact of 7)_____ on motion planning and reconfiguration.
	F. Limitations and Weaknesses of the High-Level Planner	Explores the limitation of the high-level planner due to 8)_____ about component reliability, and highlights the weakness of 9)_____ in centralized, hierarchical systems.
	G. Challenges of Open-Loop Behaviors and Robustness Enhancement	Addresses the 10)_____ of open-loop behaviors to hardware errors and environmental 11)_____, proposing solutions to enhance system robustness through closed-loop sensing and 12)_____.

Continued Table

Function	Paragraph	Main Content
Conclusion	H. Limitations and Improvement of Perception Characterization	Discusses the limitations and 13)_____ of the perception characterization component, advocating the need for a more general characterization function to handle complex environments.

Rise of a Great Power

Translate the following Chinese text into English.

The academic community, research institutions, and industry in China 1)(积极合作,致力于设计可变形、可重组的模块化机器人系统,以实现任务的灵活性和自适应性) _____

_____.

Furthermore, 2)(中国在计算机视觉、激光雷达(LiDAR)和传感器等技术领域的持续发展) _____

_____

has provided more accurate and efficient environmental perception and target recognition capabilities for robot systems, 3)(为自主模块化机器人系统的应用奠定了基础) _____

_____. Through centralized control and decision-making, 4)(自主模块化机器人系统能够根据任务需求和环境条件进行动态调整,提高任务执行的效率和灵活性) _____

_____. In various fields such as industrial automation, disaster rescue, logistics, and healthcare, autonomous modular robot systems demonstrate vast application prospects.

Overall, China has made significant progress in the research and application of autonomous modular robot systems. 5)(这种创新技术的发展将为中国在机器人领域的持续创新和发展注入新的动力) _____

_____

___ and support the realization of intelligent manufacturing and the automation needs of future society.

7-1 Unit Seven-Answer and Translation

Unit Eight Mechatronics

Warm-up

Implementing the strategy of building a strong manufacturing country conforms to the fundamental requirements of the "two centenary goals" and the "Chinese Dream". Through continuously implementing "Made in China 2025", "Made in China 2035", and "Made in China 2045", China will become a global leading manufacturing power by the centenary of the founding of the People's Republic of China. With the prosperity and strength of the manufacturing industry, it will support the great rejuvenation of the Chinese nation's "Chinese Dream". Implementing "Made in China 2025" and promoting the transformation of manufacturing from big to strong is an objective requirement to achieve stable economic growth, adjust the economic structure, and improve quality and efficiency.

Thought-provoking Questions

· What are the key strategies and initiatives outlined in "Made in China 2025" and subsequent plans like "Made in China 2035" and "Made in China 2045" to transform China into a global leading manufacturing power?

· What challenges and opportunities does China face in implementing the "Made in China 2025" and subsequent plans to achieve its goal of becoming a global leading manufacturing power?

Text A
The Versatility of Mechatronics Engineering

Research Background

Mechatronics engineering is an interdisciplinary field that integrates mechanical, electrical, electronic, and computer science principles to design and develop complex systems. Initially combining mechanics and electronics, mechatronics has evolved to include various technical areas, and it plays a crucial role in advanced automated industries. The term "mechatronics" originated in Japan and has now become a globally recognized term in the field of engineering. With advancements in information technology and computational intelligence, mechatronics has experienced significant growth and revolutionized various industries.

Introduction Questions

· What is mechatronics engineering, and what disciplines does it encompass?

· How has mechatronics evolved over time, and what led to its broadened definition?

· What are the applications and significance of mechatronics engineering in various industries?

A. Mechatronics engineering also called mechatronics, is an interdisciplinary branch of engineering that focuses on the integration of mechanical, electrical and electronic engineering systems, and also includes a combination of robotics, electronics, computer science, telecommunications, systems, control, and product engineering.

B. As technology advances over time, various subfields of engineering have succeeded in both adapting and multiplying. The intention of mechatronics is to produce a design solution that unifies each of these various subfields. Originally, the field of mechatronics was intended to be nothing more than a combination of mechanics, electrical and electronics, hence the name being a portmanteau of the words "mechanics" and "electronics"; however, as the complexity of technical systems continued to evolve, the definition had been broadened to include more technical areas.

C. The word mechatronics originated in Japanese-English and was created by Tetsuro Mori, an engineer of Yaskawa Electric Corporation. It was registered as trademark by the company in Japan with the registration number of "46-32714" in 1971. The company later released the right to use the word to the public, and the word began being used globally. Currently the word is translated into many languages and is considered an essential term for advanced automated industry.

D. Many people treat mechatronics as a modern buzzword synonymous with automation, robotics and electromechanical engineering. French standard NF E 01-010 gives the following definition: "approach aiming at the synergistic integration of mechanics, electronics, control theory, and computer science within product design and manufacturing, in order to improve and/or optimize its functionality". The word mechatronics was registered as trademark by the company in Japan with the registration number of "46-32714" in 1971. The company later released the right to use the word to the public, and the word began being used globally.

E. With the advent of information technology in the 1980s, microprocessors were introduced into mechanical systems, improving performance significantly. By the 1990s, advances in computational intelligence were applied to mechatronics in ways that revolutionized the field.

F. A mechatronics engineer unites the principles of mechanics, electrical, electronics, and computing to generate a simpler, more economical and reliable system. Engineering cybernetics deals with the question of control engineering of mechatronic systems. It is used to control or regulate such a system (control theory). Through collaboration, the mechatronic modules perform the production goals and inherit flexible and agile manufacturing properties in the production scheme. Modern production equipment consists of mechatronic modules that are integrated according to a control architecture. The most known architectures involve hierarchy, polyarchy, heterarchy, and hybrid. The methods for achieving a technical effect are described by control algorithms, which might or might not utilize formal methods in their design. Hybrid systems important to mechatronics include production systems, synergy drives, exploration rovers, automotive subsystems such as anti-lock braking systems and spin-assist, and everyday equipment such as autofocus cameras, video, hard disks, CD players and phones.

G. Mechanical engineering is an important part of mechatronics engineering. It includes the study of mechanical nature of how an object works. Mechanical elements refer to mechanical structure, mechanism, therm-fluid, and hydraulic aspects of a mechatronics system. The study of thermodynamics, dynamics, fluid mechanics, pneumatics and hydraulics. Mechatronics engineer who works a mechanical engineer can specialize in hydraulics and pneumatics systems, where they can be found working in automobile industries. A mechatronics engineer can also design a vehicle since they have strong mechanical and electronical background. Knowledge of software applications such as computer-aided design and computer aided manufacturing is essential for designing products. Mechatronics covers a part of mechanical syllabus which is widely applied in automobile industry.

H. Mechatronic systems represent a large part of the functions of an automobile. The control loop formed by sensor—information processing—actuator—mechanical (physical) change is found in many systems. The system size can be very different. The Anti-lock braking system (ABS) is a mechatronic system. The brake itself is also one. And the control loop formed by driving control (for example cruise control), engine, vehicle driving speed in the real world and speed measurement is a mechatronic system, too. The great importance of mechatronics for automotive engineering is also evident from the fact that vehicle manufacturers often have development departments with "Mechatronics" in their names.

I. Electronics and Telecommunication engineering specializes in electronics devices and telecom devices of a mechatronics system. A mechatronics engineer specialized in electronics and telecommunications have knowledge of computer hardware devices. The transmission of signal is the main application of this subfield of mechatronics. Where digital and analog systems also forms an important part of mechatronics systems. Telecommunications engineering deals with the transmission of information across a medium.

J. Electronics engineering is related to computer engineering and electrical engineering. Control engineering has a wide range of electronic applications from the flight and propulsion systems of commercial airplanes to the cruise control present in many modern cars. VLSI designing is important for creating integrated circuits. Mechatronics engineers have deep knowledge of microprocessors, microcontrollers, microchips and semiconductors. The application of mechatronics in electronics manufacturing industry can conduct research and development on consumer electronic devices such as mobile phones, computers, cameras etc. For mechatronics engineers it is necessary to learn operating computer applications such as MATLAB and Simulink for designing and developing electronic products.

K. In summary, mechatronics engineering is an interdisciplinary field that integrates mechanical, electrical, and electronic engineering systems. It combines principles from various technical areas and plays a crucial role in improving functionality and optimizing product design and manufacturing. Mechatronics has revolutionized industries such as automotive engineering, electronics manufacturing, and telecommunications. It offers a comprehensive understanding of both electrical and mechanical systems, making it a vital discipline in today's technological advancements.

Vocabulary List

mechatronics

the combination of mechanical and electronic systems　机电一体化

telecommunications

the transmission of information over long distances using electronic systems　电信,远程通信

portmanteau

a word blending the sounds and meanings of two others　混成词

synergistic

achieving greater effectiveness or efficiency through combined action　协同的,协同作用的

hierarchy

a system in which individuals or groups are ranked according to status or authority　层级,等级制度

polyarchy

a form of government in which power is vested in multiple centers　多元政权

heterarchy

a system of organization where elements have equal or similar authority　异等组织

therm-fluid

relating to the study of heat and fluid dynamics　热流体

hydraulic

operated, moved, or controlled by liquid or fluid pressure   液压的

pneumatics

relating to the study or use of pressurized gas to transmit and control mechanical energy   气动学

analog

relating to or using signals or information represented by a continuously variable physical quantity   模拟的

Language Enhancement

I. Complete the following sentences with words listed in the box below. Change the form where necessary.

multiply	brake	integrate	propulsion	utilize
collaborate	subfield	architecture	trademark	optimize
synonymous	buzzword	revolutionize	agile	unify

1. _____ of various technologies has improved the efficiency of our manufacturing process.

2. In the field of computer science, there are several _____ such as artificial intelligence and cybersecurity.

3. By using advanced algorithms, we can _____ the processing speed of our computer systems.

4. The goal of our project is to _____ different software platforms into a single, user-friendly interface.

5. Our company successfully registered a _____ for our innovative product.

6. "Big data" has become a _____ in the tech industry, often synonymous with advanced analytics.

7. We continuously _____ our website to enhance the user experience and increase conversion rates.

8. The invention of smartphones _____ the way we communicate and access information.

9. Effective _____ among team members is crucial for the success of any tech project.

10. _____ methodologies allow for quick adaptation to changing requirements in software development.

11. The _____ of our network infrastructure is designed to handle high volumes of data traffic.

12. We _____ machine learning algorithms to analyze customer behavior and make personalized recommendations.

13. The autonomous vehicle's advanced _____ system ensures quick and precise stopping.

14. Electric _____ is a promising technology for reducing emissions in the transportation sector.

15. Artificial intelligence and machine learning have become almost _____ terms in the field of data analysis and predictive modeling.

Ⅱ. Complete the following sentences with phrases listed in the box below. Change the form where necessary.

| specialize in | engage in | nothing more than |
| with the advent of | be intended to | |

1. _____ cutting-edge technology, the new software is intended to be a game-changer in the field of artificial intelligence.

2. In the world of cybersecurity, some hackers are _____ opportunistic individuals who engage in illegal activities for personal gain.

3. The research institute _____ developing innovative solutions for renewable energy, aiming to reduce carbon emissions and combat climate change.

4. The company _____ a pioneer in the development of autonomous vehicles, engaging in extensive testing and research to ensure safe and efficient transportation.

5. As a tech enthusiast, I love to _____ coding and programming projects in my free time.

Ⅲ. Complete the summary with words and phrases listed in the box. Change the form where necessary.

include	reliable	encompass	merge	integrate
possess	transmission	specialize in	combine	complex
comprehensive	focus on	original	incorporate	revolutionize

Mechatronics engineering is an interdisciplinary field that 1)_____ mechanical, electrical, and electronic engineering systems. It 2)_____ robotics, electronics, computer science, telecommunications, control, and product engineering. 3)_____, mechatronics aimed to 4)_____ mechanics, electronics, and electrical engineering, but as technical systems became more 5)_____, the definition expanded to include additional technical areas.

The term "mechatronics" was coined by Tetsuro Mori of Yaskawa Electric Corporation and registered as a trademark in Japan. With the advent of information technology, microprocessors were 6)_____ into mechanical systems, 7)_____ the field. Mechatronics engineers unite principles from various disciplines to create simpler, more 8)_____ systems.

Mechanical engineering is a crucial component of mechatronics, 9)_____ the study of mechanical structures, mechanisms, thermodynamics, fluid mechanics, pneumatics, and hydraulics. Mechatronics engineers can 10)_____ areas like hydraulics and pneumatics, working in industries such as automobiles. Knowledge of software applications and computer-aided design is also essential.

Mechatronic systems play a significant role in automobiles, 11)_____ functions like anti-lock braking systems and driving control. Electronics and telecommunication engineering within mechatronics 12)_____ devices and signal 13)_____, while electronics engineering is connected to computer engineering and control engineering. Mechatronics engineers 14)_____ expertise in microprocessors, microcontrollers, and semiconductors.

Overall, mechatronics engineering is a 15)_____ discipline that combines various fields to design and develop innovative systems for diverse industries.

Academic Expression

Pair work: Discuss with your partner and compare the two possible paraphrases of each sentence and decide which one expresses the original meaning more academically.

1. Mechatronics engineering is an interdisciplinary branch of engineering that focuses on the integration of mechanical, electrical and electronic engineering systems, and also includes a combination of robotics, electronics, computer science, telecommunications, systems, control, and product engineering.

    a. Mechatronics engineering is an interdisciplinary field that emphasizes the integration of mechanical, electrical, and electronic engineering systems, encompassing robotics, electronics, computer science, telecommunications, systems engineering, control engineering, and product engineering.

  b. Mechatronics engineering is a mix of different branches of engineering that combines mechanical, electrical, and electronic systems. It includes things like robotics, electronics, computer science, telecommunications, and designing products.

2. With the advent of technology advances over time, various subfields of engineering have succeeded in both adapting and multiplying.

  a. As technology has gotten better and better, different branches of engineering have found ways to adjust and grow.

  b. As technology has advanced, different subfields of engineering have successfully adapted and proliferated.

3. Originally, the field of mechatronics was intended to be nothing more than a combination of mechanics, electrical and electronics, hence the name being a portmanteau of the words "mechanics" and "electronics".

  a. At first, mechatronics was just a combination of mechanics, electrical engineering, and electronics, that's why they called it that.

  b. Initially, mechatronics was simply meant to be a fusion of mechanics, electrical engineering, and electronics, hence the name combining "mechanics" and "electronics".

4. The word mechatronics originated in Japanese-English and was created by Tetsuro Mori, an engineer of Yaskawa Electric Corporation. It was registered as trademark by the company in Japan with the registration number of "46-32714" in 1971. The company later released the right to use the word to the public, and the word began being used globally. Currently the word is translated into many languages and is considered an essential term for advanced automated industry.

  a. The term "mechatronics" was coined by Tetsuro Mori, an engineer at Yaskawa Electric Corporation, in Japanese-English. It was registered as a trademark by the company in Japan with the registration number "46-32714" in 1971. The company subsequently granted public usage rights for the term, leading to its global adoption. Presently, the term has been translated into numerous languages and is recognized as a fundamental concept in advanced automation industry.

  b. The word "mechatronics" was made up by an engineer named Tetsuro Mori at Yaskawa Electric Corporation. They trademarked the word in Japan back in 1971, but later let other people use it too. Now, It's used all over the world and is an important term in advanced automation.

5. Many people treat mechatronics as a modern buzzword synonymous with automation, robotics and electromechanical engineering.

　　a. A lot of people think mechatronics is just a fancy word for automation, robotics, and electromechanical engineering.

　　b. Mechatronics is often perceived as a contemporary buzzword that is equivalent to automation, robotics, and electromechanical engineering.

6. With the advent of information technology in the 1980s, microprocessors were introduced into mechanical systems, improving performance significantly. By the 1990s, advances in computational intelligence were applied to mechatronics in ways that revolutionized the field.

　　a. The introduction of information technology in the 1980s led to the integration of microprocessors into mechanical systems, resulting in significant performance enhancements. In the 1990s, the application of computational intelligence in mechatronics brought about revolutionary changes in the field.

　　b. In the 1980s, they started putting microprocessors into mechanical systems thanks to information technology, and it made them work a whole lot better. Then in the 1990s, they used smart computer stuff in mechatronics and totally changed the field.

Understanding the Text

Ⅰ. **Pair work: Work with your partner and take turns asking and answering the following questions according to the information contained in the text.**

1. What is mechatronics engineering?

2. How has the definition of mechatronics evolved over time?

3. Who coined the term "mechatronics" and when?

4. How has mechatronics become a global term?

5. How is mechatronics related to automation and robotics?

6. How did information technology impact mechatronics?

7. What are the main components of mechatronics engineering?

___

8. How does mechatronics apply to industries such as automotive and electronics?

___

Ⅱ. Group work: Please work in groups and complete the structural table with the information from Text A.

Section	Paragraph	Content
Introduction	A	Definition of mechatronics engineering as an 1)_____ branch that 2)_____ mechanical, electrical, and electronic engineering systems, including robotics, computer science, and telecommunications.
Main Body	B	3)_____ of mechatronics, originally combining mechanics, electrical, and electronics, but 4)_____ to include more technical areas.
	C	5)_____ of the word "mechatronics" in Japanese-English and its global usage in advanced automated industry.
	D	Mechatronics 6)_____ as the integration of mechanics, electronics, control theory, and computer science in product design and manufacturing.
	E	7)_____ in information technology introducing microprocessors to improve performance in mechatronic systems.
	F	8)_____ of mechatronics engineer in uniting mechanics, electrical, electronics, and computing, and application areas.
	G	9)_____ between mechanical engineering and mechatronics, application in automobile industry, and software design.
	H	10)_____ of mechatronics in automotive engineering and examples of mechatronic systems in vehicles.
	I	11)_____ in electronics and telecommunications engineering within mechatronics, signal transmission, and digital systems.
	J	12)_____ of mechatronics in electronics manufacturing, control engineering in electronic systems, and use of computer applications.

Continued Table

Section	Paragraph	Content
Conclusion	K	Mechatronics engineering integrating mechanical, electrical, and electronic engineering systems to 13) _____ functionality and advance technological applications.

Rise of a Great Power

Translate the following Chinese part into English and make it a complete English text.

Mechatronics engineering in China is a rapidly developing field 1)(它将机械工程、电气工程和控制工程相结合,形成了一个跨学科的领域) _____

_____

_____. By this integration, 2)(机电一体化工程可以提高生产效率、降低成本、提升产品质量和性能) _____

_____

_____. For instance, in the manufacturing industry, the application of mechatronics systems enables automated production lines, 3)(提高生产效率和产品一致性) _____

_____. In the healthcare sector, mechatronics technology 4)(机电一体化技术可以帮助开展精确的手术和诊断,提高医疗水平) _____

_____

_____.

Through continuous efforts, China's mechatronics engineering will continue to grow and 5)(为实现智能制造和高效能源利用等目标做出重要贡献) _____

_____

_____.

Text B
Intelligent Manufacturing Execution Systems

Research Background

The concept of Manufacturing Execution System (MES) emerged in the mid-1990s to address the limitations of the Enterprise Resource Planning (ERP) layer in real-time management of operations on the shop floor. While ERP modules focus on planning, inventory control, and accounting, they lack the speed and level of detail required to respond immediately to transactions on the shop floor. To bridge this gap, MES was developed to establish a connection between the shop floor and the ERP layer, enabling real-time management of production activities.

Introduction Questions

· What challenges did the emergence of Manufacturing Execution System (MES) aim to address?

· What are the main functionalities of Manufacturing Execution System (MES) and how do they benefit manufacturing enterprises?

· How does Manufacturing Execution System (MES) contribute to the realization of Industry 4.0?

A. The concept of Manufacturing Execution System (MES) emerged in mid-1990s to address the Enterprise Resource Planning (ERP) layer's insufficiency in real-time management of operations on the shop floor. ERP includes modules for production planning, inventory control, demand forecasting, cost accounting, and marketing for manufacturing enterprises. Since ERP collects and integrates the information required for implementing these modules from the shop floor and other organizational functions on a daily, weekly, or monthly basis, it lacks the speed and level of detail that is vital to respond immediately to every single transaction occurring on the shop floor.

B. To solve this issue, the MES concept was developed to make a connection between the shop floor and ERP layer. From a top-down view of the management hierarchy, MES creates a detailed operational plan by combining preliminary production plans from the ERP with real-time information on processes, materials, and operations from the machines, controls, and individuals on the shop floor. This plan enables real-time management of production activities on the shop floor from order receipt to finished products. From a bottom-up view, MES provides ERP with abstract information on the shop floor execution. For instance, it updates ERP on the completion status of an order, which can affect the release of upcoming planned orders.

C. The main functionalities of the MES are data acquisition and abstraction, detailed scheduling of operations, resource allocation and control, dispatching production to machines and workers, controlling product quality, and managing the maintenance of equipment and tools. The implementation of MES in manufacturing enterprises improves the critical key performance indicators (KPIs) of the company including reduced lead time and cost, improved quality, production transparency, and increased efficiency. In addition, MES can provide online information on the raw material inventory, machine breakdowns, and delays at the shop floor of each manufacturer across the supply chain. Thus, it also helps the Supply Chain Management (SCM) layer, which interacts with the ERP layer, to better respond to disturbances and disruptive events.

D. Industry 4.0 is defined as bringing intelligence, flexibility, operational efficiency, and fully predictive production to manufacturing enterprises. The first step toward achieving this goal is to collect and integrate data from the MES as well as other information systems, the internet, and manufacturing resources. The collected data can then be analyzed and used to bring intelligence to the product design, planning, and production stages, as well as equipment maintenance. In particular, MES can be realized as one of the key enablers of the fourth industrial revolution in manufacturing enterprises due to two important reasons. First and foremost, the fundamental features of MES serve as the foundation for implementing Industry 4.0 concepts. Second, using Cyber-Physical Systems (CPS) and Cyber-Physical Production Systems (CPPS), MES can enable business processes in ERP and tiers across the supply chain to become smarter by supplying online data from smart products and machines on the shop floor.

E. Integrating Artificial Intelligence (AI) with MES is one of the main research frontiers with the goal of adopting current generation of MES to the Industry 4.0 context. Production lines consist of robots, conveyor belts, machines, and supporting activities such as maintenance, quality control, and material handling aiming to efficiently manufacture the desired product. The inherent inter-dependency and uncertainty in manufacturing operations lead to a non-linear and stochastic system. Despite this complexity, such systems should operate in an optimal condition in order to keep the company competitive in the market and meet the productivity, quality, and cost objectives, while guaranteeing safety in the working environment. AI tools have the unique capability to classify and identify non-linear and multivariate patterns that remain undiscovered by the production engineers.

F. Moreover, AI tools such as machine learning and deep learning algorithms can be trained by big data sets generated by machines, ambient sensors, controllers, and worker records to reveal patterns that can contain important clues to solve challenging problems in manufacturing processes. AI tools are widely applied to MES for productivity estimation, quality faults detection, root cause diagnosis of quality defects, job dispatching and scheduling, resource allocation, human-robot collaboration, machine vision, robot manipulation, condition-based/ predictive maintenance, and manufacturing process control.

G. Digital Twin (DT) and Augmented Reality (AR) are two other current research frontiers in MES. DT is a simulation model representing the same characteristics of a physical object in the digital world. In other words, DT creates a bilateral connection between the intelligence layer of MES and the physical world. It obtains data from the production line by reading sensors embedded in the shop floor in real-time. The current state of the production line is then provides to the MES intelligence layer. The intelligence layer uses AI tools to automatically supervise and control the physical world based on the inputs provided by the DT. After decision making, the intelligence layer provides the DT with actions that should be taken on the physical equipment. AR is a core technology in facilitating human integration with MES.

H. Although the focus of industry 4.0 is to build MESs that are seamlessly intelligent, flexible, adaptable, and autonomous, humans still play a key role in industry 4.0-based MESs. AR is an interface through which humans can communicate with the digital world of the production line. Assembly, maintenance, quality inspection, and logistics activities are the main fields in manufacturing systems where AR can be applied to support human operators and provide them with visual data.

Vocabulary List

preliminary
prior to or preparing for the main matter, action, or event 初步的;预备的;开始的

transparency

the condition of being transparent or open to scrutiny　透明度;公开;透明性

enabler

something or someone that enables or makes possible　使能者;促进者;助推器

stochastic

randomly determined or involving a random variable　随机的;概率的;统计的

ambient

relating to the immediate surroundings or environment　环境的;周围的;氛围的

Language Enhancement

Ⅰ. Complete the following sentences with words listed in the box below. Change the form where necessary.

transparent	disruptive	interface	diagnosis	insufficient
disturb	bilateral	inherent	transaction	execution
defect	predictive	preliminary	allocation	dispatch

1. The _____ of resources hindered the completion of the project on time.

2. The financial _____ was securely processed through an encrypted online payment gateway.

3. Before starting the experiment, the researchers conducted _____ tests to ensure accurate results.

4. The _____ of the software program was flawless, resulting in efficient and reliable performance.

5. The _____ of funds for research and development projects was based on their potential impact and feasibility.

6. The team quickly _____ technicians to resolve the technical issue reported by the customer.

7. The company emphasizes _____ by providing detailed reports on its financial activities to shareholders.

8. The unexpected _____ in the power supply caused a temporary shutdown of the manufacturing process.

9. The _____ technology revolutionized the market, rendering traditional methods obsolete.

10. Using _____ analytics, the system accurately forecasted customer demand and optimized inventory levels.

11. The _____ quality of the product ensured its durability and reliability.

12. The doctor conducted a thorough _____ to identify the cause of the patient's symptoms.

13. The _____ in the manufacturing process resulted in a batch of faulty products.

14. The _____ agreement between the two countries promoted cooperation in the field of renewable energy.

15. The user-friendly _____ of the software made it easy for customers to navigate and utilize its features.

Ⅱ. Complete the summary with the proper words or phrases in Text B.

The concept of Manufacturing Execution System (MES) emerged in the mid-1990s to address the insufficiency of 1) _____ management of operations on the shop floor in Enterprise Resource Planning (ERP) systems. MES connects the shop floor with the ERP layer, creating a detailed operational plan by combining 2) _____ production plans with real-time information. It provides functionalities such as 3) _____, scheduling, resource allocation, 4) _____, and maintenance management. Implementation of MES improves key performance indicators, enhances production 5) _____, and helps the Supply Chain Management layer respond to 6) _____. MES plays a crucial role in achieving Industry 4.0 goals by integrating with Cyber-Physical Systems, enabling smarter business processes. Integrating Artificial Intelligence (AI) with MES enhances 7) _____, quality control, 8) _____, and maintenance. Digital Twin (DT) and Augmented Reality (AR) are current research frontiers in MES, enabling simulation and human integration. While industry 4.0 emphasizes 9) _____, humans still play a key role in MES, with AR facilitating 10) _____ between operators and the digital world.

Academic Expression

Pair work: Discuss with your partner and compare the two possible paraphrases of each sentence and decide which one expresses the original meaning more academically.

1. The concept of Manufacturing Execution System (MES) emerged in the mid-1990s to

address the Enterprise Resource Planning (ERP) layer's insufficiency in real-time management of operations on the shop floor.

  a. Manufacturing Execution System (MES) was developed in the mid-1990s to improve the real-time management of operations on the shop floor, which was lacking in the Enterprise Resource Planning (ERP) system.

  b. The emergence of Manufacturing Execution System (MES) took place in the mid-1990s to mitigate the deficiency of real-time operation management on the shop floor within the Enterprise Resource Planning (ERP) layer.

2. To solve this issue, the MES concept was developed to make a connection between the shop floor and ERP layer.

  a. To fix this problem, they came up with the MES concept to connect the shop floor and the ERP layer.

  b. To resolve this problem, the MES concept was devised to establish a linkage between the shop floor and the ERP layer.

3. The main functionalities of the MES are data acquisition and abstraction, detailed scheduling of operations, resource allocation and control, dispatching production to machines and workers, controlling product quality, and managing the maintenance of equipment and tools.

  a. The primary functionalities of the MES encompass data acquisition and abstraction, comprehensive operational scheduling, resource allocation and control, production distribution to machines and workers, product quality control, and equipment and tool maintenance management.

  b. The main things that the MES does include collecting and organizing data, making detailed schedules for operations, assigning resources and controlling them, sending production to machines and workers, checking product quality, and managing equipment and tool maintenance.

4. The implementation of MES in manufacturing enterprises improves the critical key performance indicators (KPIs) of the company including reduced lead time and cost, improved quality, production transparency, and increased efficiency.

  a. When they put MES into action in manufacturing companies, it makes the important indicators like time and cost go down, improves the quality, makes production easier to understand, and makes things run more efficiently.

  b. The implementation of MES in manufacturing enterprises enhances the crucial key performance indicators (KPIs) of the company, encompassing decreased lead time and cost, enhanced quality, improved production transparency, and heightened efficiency.

5. Integrating Artificial Intelligence (AI) with MES is one of the main research frontiers with the goal of adopting the current generation of MES to the Industry 4.0 context.

 a. The integration of Artificial Intelligence (AI) with MES stands as one of the principal research frontiers, aiming to assimilate the present generation of MES into the Industry 4.0 context.

 b. One of the main areas they're researching is how to combine Artificial Intelligence (AI) with MES, so that they can make the current MES systems work better with the Industry 4.0 stuff.

6. Digital Twin (DT) and Augmented Reality (AR) are two other current research frontiers in MES.

 a. Digital Twin (DT) and Augmented Reality (AR) represent two additional current research frontiers within the domain of MES.

 b. They're also looking into Digital Twin (DT) and Augmented Reality (AR) as new areas to explore within MES.

Understanding the Text

I. Text B has eight paragraphs, A-H. Which paragraph contains the following information? You may use any letter more than once.

1. AI tools possess the ability to discern intricate patterns and interconnections within manufacturing operations, often overlooked by human production engineers.

2. MES's adoption in manufacturing enterprises brings about significant enhancements in key performance indicators.

3. The Industry 4.0 paradigm seeks to transform manufacturing enterprises through the integration of intelligence, flexibility, operational efficiency, and predictive production.

4. AR acts as an interface through which humans can interact with the digital production line, providing support in various tasks.

5. The MES concept was introduced to bridge the gap between the shop floor and the ERP layer.

6. The integration of Artificial Intelligence (AI) with MES is a prominent area of research, aiming to align the existing MES systems with the requirements of Industry 4.0.

7. The emergence of MES aimed to tackle the real-time operational management limitations of the shop floor in the Enterprise Resource Planning (ERP) layer.

8. AI tools like machine learning and deep learning algorithms can be trained using vast datasets generated by machines, sensors, controllers, and worker records.

9. Digital Twin (DT) and Augmented Reality (AR) are emerging areas of research in the MES field.

10. The initial phase involves gathering and integrating data from MES, other information systems, the internet, and manufacturing resources.

II. Text B has eight paragraphs, A-H. Choose the correct heading for paragraphs A-H from the list of headings below.

List of headings

I	Historical Development of Manufacturing Execution Systems
II	AI Tools for Enhancing MES in Manufacturing Processes
III	The Role of Humans in Industry 4.0-based MES Systems
IV	Industry 4.0 and the Role of MES
V	Digital Twin (DT) and its Connection with MES
VI	Integrating Artificial Intelligence (AI) with MES
VII	Challenges and Limitations of MES Implementation
VIII	Addressing the Real-time Management Limitations of ERP
IX	Introduction to Manufacturing Execution System (MES)
X	Functionalities of MES in Manufacturing Enterprises

Paragraph A:
Paragraph B:
Paragraph C:
Paragraph D:
Paragraph E:
Paragraph F:
Paragraph G:
Paragraph H:

Ⅲ. Group work: To analyze the article's structure in table form, we can break it down into sections and provide a brief description of each section. Please work in groups and complete the structural table with the information from Text B.

Section	Paragraph	Content
Background and Overview	A	Introduces the 1)_____ and 2)_____ background of MES, along with its relationship with ERP.
	B	Describes the role and 3)_____ of MES.
Functionality and Impact of MES	C	Provides detailed explanations of MES's main 4)_____ and its 5)_____ on key performance indicators and 6)_____ management.
	D	Discusses the 7)_____ role of MES in achieving Industry 4.0 goals.
Emerging Technologies and Future Development	E	Discusses the importance of 8)_____ new technologies such as artificial intelligence, digital twin, and augmented reality with MES and their 9)_____ for the manufacturing industry.
	H	Emphasizes the continued significance of human 10)_____ in Industry 4.0.

Rise of a Great Power

Translate the following Chinese part into English and make it a complete English text.

1. (随着中国制造业的转型升级和技术创新的推动)_____

_____

_____, an increasing number of enterprises are realizing the importance of adopting Manufacturing Execution Systems (MES).

In China, 2)(智能制造执行系统的应用范围非常广泛)_____

_____

_____, covering areas such as production planning, material management, equipment control, and quality management. By utilizing real-time data collection and analysis, MES can provide accurate production information

and guidance, 3)(帮助企业及时做出决策和调整生产计划)_____
_____
_____. However, the development of intelligent MES in China also faces several challenges. Firstly, there are technological challenges, 4)(包括数据安全、设备兼容性和系统集成等问题)_____
_____
_____. Secondly, there are challenges in talent cultivation and management, 5)(需要培养更多具备智能制造技术和管理能力的人才)_____
_____
_____.

With continuous technological advancements and government support, it is believed that China's intelligent manufacturing field will achieve even greater accomplishments and make significant contributions to the sustainable development of China's manufacturing industry.

8-1 Unit Eight-Answer and Translation

Unit Nine　Control Theory

Warm-up

To realize the shift in China's economic development without losing speed, and promote the industrial structure to move towards the middle and high end, the key, difficult points and way out are all in the manufacturing industry. The formulation of the "Made in China 2025" strategic plan is to cope with the profound impact of this series of changes, aiming at key links such as innovation-driven, intelligent transformation, strengthening the foundation, and green development, and promoting the manufacturing industry from big to strong.

Thought-provoking Questions

· As a mechanical engineering student, how can your knowledge and skills contribute to the goals of the "Made in China 2025" plan, particularly in terms of innovation-driven and intelligent transformation in the manufacturing industry?

· In the context of promoting the manufacturing industry from big to strong, what challenges and opportunities do you foresee as a future professional in the field of mechanical manufacturing?

· How can you actively participate in addressing these key points and contribute to the industry's growth and development?

Text A
Evolving Control Theory: From Engineering to AI

Research Background

Control theory is a field of control engineering that focuses on managing dynamical systems. It aims to develop models and algorithms for controlling system inputs to achieve desired states while minimizing errors and ensuring stability. Control theory has revolutionized industries like manufacturing, aircraft, and robotics by enabling precise control of processes. It has also expanded into diverse fields such as economics and artificial intelligence, where it plays a crucial role in market control and human interaction modeling.

Introduction Questions

· What is the objective of control theory and how does it contribute to the field of engineering?

· What are the key components and principles of control theory?

· How has control theory expanded its applications beyond traditional engineering disciplines?

A. Control theory is a field of control engineering and applied mathematics that deals with the control of dynamical systems in engineered processes and machines. The objective is to develop a model or algorithm governing the application of system inputs to drive the system to a desired state, while minimizing any delay, overshoot, or steady-state error and ensuring a level of control stability; often with the aim to achieve a degree of optimality.

B. To do this, a controller with the requisite corrective behavior is required. This controller monitors the controlled process variable (PV), and compares it with the reference or set point (SP). The difference between actual and desired value of the process variable, called the error signal, or SP-PV error, is applied as feedback to generate a control action to bring the controlled process variable to the same value as the set point. Other aspects which are also studied are control ability and observability. Control theory is used in control system engineering to design automation that have revolutionized manufacturing, aircraft, communications and other industries, and created new fields such as robotics.

C. Extensive use is usually made of a diagrammatic style known as the block diagram. In it the transfer function, also known as the system function or network function, is a mathematical model of the relation between the input and output based on the differential equations describing the system.

D. Control theory dates from the 19th century, when the theoretical basis for the operation of governors was first described by James Clerk Maxwell. Control theory was further advanced by Edward Routh in 1874, Charles Sturm in 1895, and Adolf Hurwitz, who all contributed to the establishment of control stability criteria; and from 1922 onwards, the development of PID control theory by Nicolas Minorsky. Although a major application of mathematical control theory is in control systems engineering, which deals with the design of process control systems for industry, other applications range far beyond this. As the general theory of feedback systems, control theory is useful wherever feedback occurs—thus control theory also has applications in life sciences, computer engineering, sociology and operations research.

E. Although control systems of various types date back to antiquity, a more formal analysis of the field began with a dynamics analysis of the centrifugal governor, conducted by the physicist James Clerk Maxwell in 1868, entitled On Governors. A centrifugal governor was already used to regulate the velocity of windmills. Maxwell described and analyzed the phenomenon of self-oscillation, in which lags in the system may lead to overcompensation and unstable behavior. This generated a flurry of interest in the topic, during which Maxwell's classmate, Edward John Routh, abstracted Maxwell's results for the general class of linear systems. Independently, Adolf Hurwitz analyzed system stability using differential equations in 1877, resulting in what is now known as the Routh-Hurwitz theorem.

F. A notable application of dynamic control was in the area of crewed flight. The Wright brothers made their first successful test flights on December 17, 1903, and were distinguished by their ability to control their flights for substantial periods (more than the ability to produce lift from an airfoil, which was known). Continuous, reliable control of the airplane was necessary for flights lasting longer than a few seconds.

G. By World War II, control theory was becoming an important area of research. Irmgard Flügge-Lotz developed the theory of discontinuous automatic control systems, and applied the bang-bang principle to the development of automatic flight control equipment for aircraft. Other areas of application for discontinuous controls included fire-control systems, guidance systems and electronics.

H. Sometimes, mechanical methods are used to improve the stability of systems. For example, ship stabilizers are fins mounted beneath the waterline and emerging laterally. In contemporary vessels, they may be gyroscopically controlled active fins, which have the capacity to change their angle of attack to counteract roll caused by wind or waves acting on the ship.

I. Control theory has also seen an increasing use in fields such as economics and artificial intelligence. Here, one might say that the goal is to find an internal model that obeys the good regulator theorem. So, for example, in economics, the more accurately a (stock or commodities) trading model represents the actions of the market, the more easily it can control that market (and extract "useful work" (profits) from it). In AI, an example might be a chatbot modeling the discourse state of humans: the more accurately it can model the human state (e. g. on a telephone voice-support hotline), the better it can manipulate the human (e. g. into performing the corrective actions to resolve the problem that caused the phone call to the help-line). These last two examples take the narrow historical interpretation of control theory as a set of differential equations modeling and regulating kinetic motion, and broaden it into a vast generalization of a regulator interacting with a plant.

Vocabulary List

diagrammatic
presented or represented in the form of a diagram   图解的;示意图的

centrifugal governor
a device that regulates the speed of an engine by controlling the fuel supply   离心调速器

velocity
the speed of an object in a particular direction   速度

bang-bang principle

a control strategy that switches abruptly between two states or actions　突变原理；突变控制

gyroscopically

in a way that relates to or is influenced by the properties or behavior of a gyroscope　像陀螺仪地

theorem

a statement or principle that can be proven to be true through logical reasoning　定理；原理

chatbot

a computer program designed to simulate conversation with human users　聊天机器人

Language Enhancement

Ⅰ. **Complete the following sentences with words listed in the box below. Change the form where necessary.**

discourse	range	kinetic	contemporary	distinguish
resolve	interpret	laterally	dynamical	govern
counteract	mount	extensive	optimal	overshoot

1. The _____ behavior of the system can be analyzed using mathematical models.

2. Encryption techniques _____ the security of online transactions and data privacy.

3. The rocket's trajectory _____ the target due to a miscalculation in the propulsion system.

4. The _____ of the algorithm was demonstrated by its ability to find the most efficient solution.

5. The research project involved a(n) _____ study of nanotechnology and its applications.

6. The drone has a wide _____ of capabilities, including aerial photography and package delivery.

7. It is important to _____ between virtual reality and augmented reality technologies.

8. The satellite was _____ on the spacecraft for deployment into orbit.

9. The robot moved _____ to avoid obstacles in its path.

10. The smartphone incorporates _____ features such as facial recognition and augmented reality.

11. To _____ the effects of climate change, scientists are developing renewable energy technologies.

12. The conference provided a platform for scientific _____ and knowledge sharing.

13. Engineers worked together to _____ the technical issues in the software development process.

14. The _____ of the experimental data revealed significant trends and patterns.

15. The _____ energy of the moving car was converted into electrical energy through regenerative braking.

II. Complete the following sentences with phrases listed in the box below. Change the form where necessary.

result in	date back to	be known as
contribute to	a flurry of	with the aim to

1. The release of the new smartphone model caused _____ pre-orders and excitement among tech enthusiasts.

2. Insufficient testing can _____ software bugs and system failures.

3. The groundbreaking invention of the transistor _____ a major milestone in the history of electronics.

4. The research project was initiated _____ develop a sustainable renewable energy source.

5. The collaborative efforts of scientists and engineers _____ advancements in artificial intelligence.

6. The origins of computer programming _____ the early 19th century with the development of punch-card systems.

II. Skim the text and complete the summary with words listed in the box below. Change the form where necessary.

desire	different	algorithm	expand	contribute
kinetic	optimal	observable	extend	onwards
stability	transfer	broaden	internal	antiquity

Control theory is a field that deals with the control of dynamical systems in engineered processes and machines. Its objective is to develop models and 1)_____ to drive systems to desired states while minimizing delay, overshoot, and steady-state error. The aim is to achieve 2)_____ control and 3)_____.

A controller is used to monitor the process variable and compare it with a reference or set point. The difference between the actual and 4)_____ values, known as the error signal, is used as feedback to generate control actions. Control theory also considers aspects like controllability and 5)_____.

The use of block diagrams and 6)_____ functions helps analyze the relationship between input and output based on 7)_____ equations. Control theory has a long history, with early 8)_____ by James Clerk Maxwell, Edward Routh, and Adolf Hurwitz. PID control theory, developed by Nicolas Minorsky, became significant from 1922 9)_____.

While control systems have been used since 10)_____, formal analysis began with Maxwell's study of the centrifugal governor. Applications of control theory 11)_____ to various fields, including crewed flight, where control was essential for longer flights. Control theory played a role in automatic flight control equipment, guidance systems, and electronics.

Control theory has 12)_____ to economics and artificial intelligence, where 13)_____ models and regulators are used to manipulate markets or model human behavior. This 14)_____ the application of control theory beyond differential equations and 15)_____ motion regulation.

Academic Expression

Pair work: Discuss with your partner and compare the two possible paraphrases of each sentence and decide which one expresses the original meaning more academically.

1. Control theory is used in control system engineering to design automation that has revolutionized manufacturing, aircraft, communications, and other industries, and created new fields such as robotics.

   a. Control theory is used in control system engineering to create automated systems that have completely changed how things are done in industries like manufacturing, aircraft, communications, and more. It has also given rise to exciting fields like robotics.

   b. Control theory is applied in the domain of control system engineering to develop automation systems that have brought about transformative changes in various sectors, including manufacturing, aircraft, communications, and diverse industries, as well as fostering the emergence of novel disciplines such as robotics.

2. The controller monitors the controlled process variable (PV) and compares it with the reference or set point (SP).

  a. The controller observes and tracks the controlled process variable (PV), making comparisons with the reference or set point (SP).

  b. The controller keeps an eye on the controlled process variable (PV) and checks if it matches the reference or set point (SP).

3. Control theory dates from the 19th century when the theoretical basis for the operation of governors was first described by James Clerk Maxwell.

  a. The origin of control theory can be traced back to the 19th century, when James Clerk Maxwell initially presented the theoretical framework for the functioning of governors.

  b. Control theory has been around since the 19th century when James Clerk Maxwell first came up with the theoretical basis for how governors work.

4. Control theory is used to design automation that minimizes delays, overshoots, and steady-state errors, while ensuring control stability and aiming for optimality.

  a. Control theory is employed in the design of automation systems that aim to minimize delays, overshoots, and steady-state errors, while simultaneously ensuring control stability and striving for optimality.

  b. Control theory is used to create automated systems that try to minimize delays, overshoots, and errors, while making sure everything stays stable and works as perfectly as possible.

5. The block diagram is extensively used in control theory to illustrate the relationships between inputs and outputs based on differential equations describing the system.

  a. The block diagram is a very useful tool in control theory to show how inputs and outputs are connected in a system, using differential equations to describe it.

  b. The block diagram finds wide-ranging application in control theory as it serves as a visual tool to depict the connections between inputs and outputs, relying on differential equations that describe the system.

6. Control theory is applicable not only in control system engineering but also in fields such as life sciences, computer engineering, sociology, and operations research.

  a. Control theory is not limited to control system engineering and extends its applicability to diverse domains including life sciences, computer engineering, sociology, and operations research.

b. Control theory is not just for control system engineering. It also has uses in other fields like life sciences, computer engineering, sociology, and operations research.

Understanding the Text

Ⅰ. **Pair work**: Work with your partner and take turns asking and answering the following questions according to the information contained in the text.

1. What is the objective of control theory?

2. What is the role of a controller in control theory?

3. How is control theory used in control system engineering?

4. What is the significance of the block diagram in control theory?

5. When did control theory originate, and who were some influential figures in its development?

6. In what other fields can control theory be applied?

7. What historical application of control theory was significant in crewed flight?

8. How has control theory influenced fields like economics and artificial intelligence?

Ⅱ. **Group work**: Please work in groups and complete the structural table with the information from Text A.

Function	Paragraph	Content
Introduction	A	Introduces the 1) _____ and 2) _____ areas of control theory.
Body	B	Expounds on the 3) _____ of control theory and the role of the controller.
	C	Discusses the importance of 4) _____ and 5) _____ in control theory as mathematical models.

Continued Table

Function	Paragraph	Content
Body	D	Traces the 6) _____ and development of control theory, including contributions to 7) _____ and the development of 8) _____.
	E	Highlights the application of 9) _____ in crewed flight.
	F	Describes the importance and 10) _____ of control theory after World War II.
	G	Explains the use of 11) _____, such as ship stabilizers, to 12) _____ system stability.
	H	Explores the 13) _____ of control theory in economics and artificial intelligence, emphasizing the concept of an 14) _____.
Conclusion	I	Summarizes the wide-ranging applications and 15) _____ of control theory.

Rise of a Great Power

Translate the following Chinese part into English and make it a complete English text.

The control theory in China is constantly evolving and developing, 1)(成为科学研究和工程应用领域的重要组成部分) _____

_____. Control theory involves the analysis, modeling, and control of system behavior 2)(以实现所需目标和性能) _____

_____. 3)(中国的控制理论广泛应用于多个领域) _____

_____. In the manufacturing industry, control theory improves production line efficiency and stability, optimizes process parameters, and enhances product quality and consistency. In the transportation sector, 4)(控制理论应用于交通流量控制和优化,提高交通系统效率和安全性) _____

_____. In the energy sector, control theory is used for stable control and optimization of power network and smart grids.

In the future, Chinese control theory will continue to develop and 5)(为高效能源利用、智能制造和可持续发展等目标做出重要贡献) _____

_____.

Text B
Closed-Loop Control: Achieving Desired System Output

Research Background

Closed-loop control systems play a crucial role in achieving precise regulation and stability in various applications. By incorporating feedback from the process variable and continuously comparing it with the desired value, these systems can adjust the control action to minimize deviations and maintain desired system outputs. Examples include cruise control in cars and temperature control in ovens. Closed-loop control enhances efficiency, reliability, and accuracy, leading to consistent performance and improved outcomes in industries ranging from automotive to manufacturing. Understanding the principles and benefits of closed-loop control is essential for advancing control engineering and optimizing system performance.

Introduction Questions

· What is the difference between open-loop control and closed-loop control?

· How does closed-loop control ensure precise regulation in systems?

· What are some examples of closed-loop control systems and their benefits?

A. Fundamentally, there exist two fundamental control loops: open-loop control and closed-loop (feedback) control. In open-loop control, the controller's control action is independent of the process output, also known as the controlled process variable (PV). An exemplary case of this is a central heating boiler that is exclusively controlled by a timer, meaning that heat is provided for a constant period, irrespective of the building's temperature. The control action is the timed switching on and off of the boiler, and the process variable is the building temperature, yet both are unlinked.

B. In closed loop control, the control action from the controller is dependent on feedback from the process in the form of the value of the process variable (PV). In the case of the boiler analogy, a closed loop would include a thermostat to compare the building temperature (PV) with the temperature set on the thermostat (the set point—SP). This generates a controller output to maintain the building at the desired temperature by switching the boiler on and off. A closed loop controller, therefore, has a feedback loop which ensures the controller exerts a control action to manipulate the process variable to be the same as the "reference input" or "set point". For this reason, closed loop controllers are also called feedback controllers.

C. The definition of a closed loop control system according to the British Standard Institution is "a control system possessing monitoring feedback, the deviation signal formed as a result of this feedback being used to control the action of a final control element in such a way as to tend to reduce the deviation to zero." Likewise, "A Feedback Control System is a system which tends to maintain a prescribed relationship of one system variable to another by comparing functions of these variables and using the difference as a means of control."

D. An example of a control system is a car's cruise control, which is a device designed to maintain vehicle speed at a constant desired or reference speed provided by the driver. The controller is the cruise control, the plant is the car, and the system is the car and the cruise control. The system output is the car's speed, and the control itself is the engine's throttle position which determines how much power the engine delivers.

E. A primitive way to implement cruise control is simply to lock the throttle position when the driver engages cruise control. However, if the cruise control is engaged on a stretch of non-flat road, then the car will travel slower going uphill and faster when going downhill. This type of controller is called an open-loop controller because there is no feedback; no measurement of the system output (the car's speed) is used to alter the control (the throttle position.) As a result, the controller cannot compensate for changes acting on the car, like a change in the slope of the road.

F. In a closed-loop control system, data from a sensor monitoring the car's speed (the system output) enters a controller which continuously compares the quantity representing the speed with the reference quantity representing the desired speed. The difference, called the error, determines the throttle position (the control). The result is to match the car's speed to the reference speed (maintain the desired system output). Now, when the car goes uphill, the difference between the input (the sensed speed) and the reference continuously determines the throttle position. As the sensed speed drops below the reference, the difference increases, the throttle opens, and engine power increases, speeding up the vehicle. In this way, the controller dynamically counteracts changes to the car's speed. The central idea of these control systems is the feedback loop, the controller affects the system output, which in turn is measured and fed back to the controller.

G. Another example of a closed-loop control system can be found in the field of temperature control for an oven. In this case, the controller's goal is to maintain the oven temperature at a specific set point, regardless of external factors such as the room temperature or heat loss from opening the oven door. In an open-loop control system, the oven may be set to operate at a fixed temperature for a predetermined duration. However, this approach fails to consider changes in the actual oven temperature and cannot make adjustments to maintain the desired temperature accurately.

H. By introducing closed-loop control, a temperature sensor inside the oven continuously monitors the actual temperature (PV) and compares it to the set point (SP). The difference between the two, known as the error, is used by the controller to adjust the oven's heating elements. If the temperature falls below the set point, the controller increases the heat output, and if the temperature exceeds the set point, the controller decreases the heat output. This feedback loop ensures that the oven temperature is constantly regulated, minimizing deviations from the desired temperature.

I. The feedback control system in the oven demonstrates the fundamental concept of closed-loop control, where feedback from the process (oven temperature) is used to adjust the control action (heating elements) to achieve the desired system output (maintaining the set temperature). This closed-loop approach enables precise temperature control, ensuring consistent cooking results and preventing overheating or undercooking.

J. By incorporating closed-loop control systems in various applications, such as the cruise control in cars or temperature control in ovens, precise regulation and stability can be achieved. These feedback control systems enhance efficiency, reliability, and accuracy in a wide range of industries, from automotive to manufacturing and beyond.

Vocabulary List

exemplary
serving as a desirable model; representing the best of its kind 典范的;出色的

irrespective
regardless of; without taking into account 不考虑的

thermostat
a device that regulates temperature 恒温器;温控器

throttle
a device controlling the flow of fuel or power to an engine 节流阀;油门;控制

predetermine
decide or establish in advance 预先确定;事先决定

duration
the length of time that something lasts 持续时间;时长

 ## Language Enhancement

Ⅰ. Complete the following sentences with words listed in the box below. Change the form where necessary.

deviation	manipulate	demonstrate	adjustment	compensate
exert	deviation	duration	primitive	exclusive
exceed	predetermine	exemplary	cruise	irrespective

1. The company's _____ performance in implementing cutting-edge technology has earned them numerous awards and recognition.

2. The _____ access to the latest scientific research allows the team to stay ahead of their competitors in technological advancements.

3. In the world of technology, success is _____ of gender, and anyone with the

right skills and knowledge can excel.

4. The engineers _____ great effort in designing and developing a groundbreaking prototype.

5. Researchers are working to _____ nanoparticles to create more efficient solar cells.

6. The _____ from the original plan caused delays in the project timeline.

7. The autonomous vehicle went on a _____ around the city, demonstrating its ability to navigate complex traffic scenarios.

8. Early computers had _____ interfaces compared to the sleek and intuitive designs of today's devices.

9. To _____ for the lack of resources, the team employed innovative techniques to achieve their technological goals.

10. The specifications of the product were _____, ensuring consistent quality across all batches.

11. The _____ of the software update process depends on the size of the program and the speed of the computer.

12. The _____ of the software settings improved the system's performance and resolved the compatibility issues.

13. The new smartphone's processing power _____ that of its competitors, providing users with faster and smoother experiences.

14. The _____ from the standard operating procedure led to errors in the experimental results.

15. The scientist used a series of experiments to _____ the effectiveness of the new drug in treating the disease.

Ⅱ. Skim the text and complete the summary with words and phrases listed in the box below. Change the form where necessary.

adjustment	incorporate	optimize	harness	demonstrate
regardless of	adaptation	application	maintain	rely on
enhance	accurate	exemplary	highlight	exclusive

This article delves into the concepts of open-loop and closed-loop (feedback) control systems, 1) _____ their applications in different scenarios. In open-loop control, the control action is independent of the process output, while closed-loop control 2) _____ feedback from the process to adjust the control action. An 3) _____ case of open-loop control is a centrally timed heating boiler that 4) _____ operates based on a predetermined duration, 5) _____ the building's temperature. In contrast, closed-loop control systems 6) _____ a feedback loop, such as a thermostat, to compare the process variable (e.g., building temperature) with a desired set point, enabling 7) _____ to the control action accordingly.

The article emphasizes the advantages of closed-loop control systems in 8) _____ precise regulation and stability. Examples, such as cruise control in cars and temperature control in ovens, 9) _____ the effectiveness of closed-loop control in achieving 10) _____ and consistent results. These feedback control systems find wide-ranging 11) _____ across industries, contributing to increased efficiency, reliability, and accuracy.

By 12) _____ closed-loop control systems in various applications, precise regulation and stability can be achieved. Feedback control 13) _____ performance and enables continuous 14) _____ based on system conditions, ultimately leading to improved efficiency, reliability, and consistency. From automotive systems to manufacturing processes, closed-loop control systems play a crucial role in 15) _____ operations and delivering reliable outcomes.

Academic Expression

Pair work: Discuss with your partner and compare the two possible paraphrases of each sentence and decide which one expresses the original meaning more academically.

1. Open-loop control is independent of the process output, as exemplified by a centrally timed heating boiler controlled solely by a timer.

   a. Open-loop control operates autonomously from the process output, as demonstrated by a centrally timed heating boiler exclusively regulated by a timer.

   b. Open-loop control is when the process is not affected by the output, like a heating boiler controlled by a timer.

2. Closed-loop control systems, also known as feedback control systems, rely on feedback from the process to adjust the control action.

   a. Closed-loop control systems, or feedback control systems, depend on process feedback to modulate the control action.

   b. Closed-loop control systems, or feedback control systems, use feedback from the process to adjust the control.

3. An open-loop controller, like cruise control without feedback, cannot compensate for changes in the road's slope.

a. An open-loop controller, like cruise control without feedback, can't adjust for changes in the road's slope.

b. An open-loop controller, such as feedback-less cruise control, lacks the ability to counterbalance variations in road gradient.

4. An open-loop control system in an oven cannot make adjustments to accurately maintain the desired temperature.

a. An open-loop control system implemented in an oven lacks the capability to precisely sustain the desired temperature.

b. An open-loop control system in an oven can't make precise adjustments to keep the desired temperature.

5. By introducing closed-loop control, an oven's temperature can be continuously regulated by comparing the actual temperature with the set point.

a. With closed-loop control, the oven's temperature is continuously regulated by comparing it to the desired temperature.

b. The implementation of closed-loop control allows for continuous regulation of an oven's temperature by comparing it with the set point.

6. A closed-loop control system uses feedback to adjust the control action and maintain the desired temperature in an oven.

a. In an oven, a closed-loop control system employs feedback to modulate the control action and sustain the desired temperature.

b. In an oven, closed-loop control uses feedback to adjust the control and keep the temperature just right.

Understanding the Text

Ⅰ. Choose the right answer according to Text B.

1. What is the main difference between open-loop control and closed-loop control?

A) Open-loop control relies on feedback from the process.

B) Closed-loop control is independent of the process output.

C) Open-loop control continuously adjusts the control action.

D) Closed-loop control uses feedback from the process to adjust the control action.

2. What is an example of open-loop control mentioned in the article?

A) Cruise control in cars.

B) Central heating boiler controlled by a timer.

C) Temperature control in an oven.

D) Throttle position control in a car engine.

3. What is the function of a thermostat in a closed-loop control system?

A) To compare the building temperature with the set point.

B) To continuously adjust the control action.

C) To measure and feed back the system output to the controller.

D) To maintain a constant period of heat supply.

4. How does a closed-loop control system maintain the desired temperature in an oven?

A) By operating at a fixed temperature for a predetermined duration.

B) By adjusting the heat output based on external factors.

C) By continuously comparing the actual temperature with the set point.

D) By using feedback from the timer to adjust the control action.

5. What is the main characteristic of an open-loop control system?

A) It relies on feedback from the process to adjust the control action.

B) It continuously compares the system output with the desired reference.

C) It uses a feedback loop to maintain the desired system output.

D) It operates independently of the process output and lacks feedback.

6. Which type of controller can compensate for changes in the road's slope?

A) Open-loop controller.

B) Closed-loop controller.

C) Feedback controller.

D) Timer-based controller.

7. What is the main objective of incorporating closed-loop control systems?

A) To enhance efficiency, reliability, and accuracy.

B) To minimize deviations in the desired temperature.

C) To maintain a constant period of heat supply.

D) To adjust the control action based on external factors.

8. What is the function of the feedback loop in a closed-loop control system?
A) To maintain the desired system output.
B) To continuously monitor the process output.
C) To dynamically adjust the control action.
D) To regulate the process variable to the reference input.

II. Do the following statements agree with the information given in Text B? Write your judgment.

TRUE   if the statement agrees with the information
FALSE   if the statement contradicts the information
NOT GIVEN   if there is no information on this

1. Open-loop control relies on feedback from the process to adjust the control action.
2. A central heating boiler controlled by a timer is an example of open-loop control.
3. The function of a thermostat in a closed-loop control system is to continuously adjust the control action.
4. Closed-loop control systems aim to maintain a constant period of heat supply.
5. An open-loop controller can compensate for changes in the road's slope.
6. Closed-loop control systems use a feedback loop to maintain the desired system output.
7. Closed-loop control systems can be used for temperature control in ovens and cruise control in cars.
8. The feedback loop in a closed-loop control system regulates the process variable to the reference input.

III. Group work: Please work in groups and complete the structural table with the information from Text B.

Function	Paragraph	Content
Introduction	A	Introduces the topic of open-loop and closed-loop control systems, and discusses the characteristics of open-loop control, 1) _____ by a central heating boiler controlled 2) _____ by a timer.
Body	B	Explains closed-loop control, using the 3) _____ of a thermostat in a heater to 4) _____ its reliance on feedback from the process.

Continued Table

Function	Paragraph	Content
Body	C	Quotes the definition of a closed-loop control system, emphasizing monitoring feedback and the use of 5)_____ signals.
	D	Presents an example of a control system, 6)_____ cruise control in cars.
	E	Describes the limitations of open-loop control in 7)_____ for changes in road slope.
	F	Explores the working principle of closed-loop control systems, 8)_____ the feedback loop's role in dynamically 9)_____ the control action.
	G	Provides another example of a closed-loop control system, temperature control in an oven.
	H	Explains the function of the feedback loop in 10)_____ the desired temperature in the oven.
Conclusion	I, J	Summarizes the key concepts of closed-loop control systems and highlights their importance in various industries.

Rise of a Great Power

Translate the following Chinese part into English and make it a complete English text.

1)(控制系统作为现代科技的核心,对于提高工业生产效率、优化能源利用和实现智能发挥着重要作用)_____

_____. China has made significant progress in automation technology, 2)(成为全球领先的控制系统研发和制造中心之一)_____

_____.

205

Domestic universities and research institutions in China actively engage in fundamental research on control systems, covering various aspects from control theory to algorithm design and engineering practices. 3)(这些研究成果为中国的工业自动化和智能制造提供了坚实的技术支持)_____
_____.

Numerous local companies actively participate in the research and application of control systems, continuously introducing products and solutions with independent intellectual property rights. Chinese control system products have been widely applied in sectors such as power, transportation, and manufacturing, gradually entering the international market. 4)(一些中国企业也积极参与国际标准的制定和技术合作,不断提升自身的技术水平和竞争力)_____
_____
_____.

5)(随着科技的不断进步和应用需求的不断增长,中国将继续加强对控制系统的研发应用)_____
_____,
making greater contributions to achieving intelligent manufacturing and sustainable development.

9-1　Unit Nine-Answer and Translation

与本书配套的二维码资源使用说明

　　本书部分课程及与纸质教材配套数字资源以二维码链接的形式呈现。利用手机微信扫码成功后提示微信登录，授权后进入注册页面，填写注册信息。按照提示输入手机号码，点击获取手机验证码，稍等片刻收到4位数的验证码短信，在提示位置输入验证码成功，再设置密码，选择相应专业，点击"立即注册"，注册成功。（若手机已经注册，则在"注册"页面底部选择"已有账号？立即登录"，进入"账号绑定"页面，直接输入手机号和密码登录。）接着提示输入学习码，需刮开教材封面防伪涂层，输入13位学习码（正版图书拥有的一次性使用学习码），输入正确后提示绑定成功，即可查看二维码数字资源。手机第一次登录查看资源成功以后，再次使用二维码资源时，只需在微信端扫码即可登录进入查看。（如申请二维码资源遇到问题，可联系宋焱：15827068411。）